MATH WITHOUT TEARS

MATH WITHOUT TEARS

Roy Hartkopf

New York: EMERSON BOOKS, Inc.

Second Printing, 1972

Library of Congress Catalog Card Number: 70-140437
Standard Book Number: 87523-173-X

Manufactured in the United States of America

To Cedric and Ted and Peggy

The author expresses his thanks to
Mr Graeme A. Wilson for his excellent
work in preparing the illustrations

CONTENTS

1. One Plus One Is Nothing 11
Universality and accuracy, an impossible combination—mathematics is neither—a language with maddening habits—what is one?—sheep-and-cow mathematics—the mathematics of measurement—square feet and square sheep—some basic rules—slipping in the minus—operators and "*i*" notation—its use in maps and vectors—the end of the theory of dimensions.

2. One Plus One Is 10 36
The basis of counting—eight-fingered arithmetic—binary counting and computers—table of comparisons—100 feet long—decimals, eighthmals, and binamals—0.001 of an inch.

3. Logarithms for Leisure 55
Doing things the hard way—tricks of the trade—powers and examples—the world of multiply—minuses and zeros—log tables and their uses—examples—"*i*" on the angle—infuriating proofs.

4. Abacus Automation 73
Working without work—the adding stick—logarithmic scales—conversion and other scales—slide rules—do it yourself—nomographs and how to make them—a useful sample.

5. Graphs . . . 100
The world of ratios—division which isn't—conversion ratios—decimal currency—cause-and-effect ratios—deaths and doctors—facts and opinions.

6. . . . AND RACKETS 114
"Simple" records—pictures that mislead—a bedtime story in graphs—the study of mankind.

7. THE RATE OF CHANGE 126
Change and time—more things and ideas—steepness and slopes—what is slope?—the rate of change—the changing rate of change—acceleration—having a successful automobile accident—another useful nomograph—some relationships of change.

8. THE RIGHT ANGLE ON TRIGONOMETRY 156
Right angles and triangles—essential facts—heights without climbing—simplified surveying—trigonometrical ratios—their names and meaning—more about slopes—negative slopes—putting direction into sides and angles—sine waves and their importance—what about "*i*"—the consistency of mathematics.

9. HIGHWAY TO NOWHERE 194
The frog that jumped—sums to infinity—repeating decimals—evaluating π—speaking exponentially.

10. CALCULUS AND CONTRADICTIONS 212
Imaginary motion—nothing divided by nothing—throwing a stone—a proposed business investment—unchanging types of change.

A HEARTENING NOTE

One of the most disheartening aspects of the problem of learning is the feeling one gets of being "out of it." The teachers, professionals, and textbooks unconsciously form a clique with a language which they alone understand. The plumber talks of "tees" and "elbows"; the builder talks of "studs" and "plates" and "architraves"; the mathematician talks of "ratios" and "logarithms" and "operators."

The beginner—especially if he is an adult—feels ignorant and inferior and is strongly tempted to retreat to his own world where he will be among familiar surroundings and will be treated as an equal.

The purpose of this book is not to train mathematicians, but to give ordinary non-mathematical mortals some idea of what it is all about. Because of this it has been possible, to a very great extent, to dispense with mathematical terms and jargon. Even so, confusion can often arise over details which when understood seem ridiculously simple. For instance the fraction one-half can be written $\frac{1}{2}$, or if the typewriter has no fractions, 1/2, or in decimals as 0.5 or plain .5. A person who doesn't realize that all these mean exactly the same may spend endless time wondering what deep mathematical meaning lies in the fact that it is written $\frac{1}{2}$ in one place and 0.5 in another.

So if you can follow the general train of thought don't worry about such details. Leave them till later, or to the unfortunate people who are studying to pass an examination!

1. ONE PLUS ONE IS NOTHING

The only point on which all philosophers agree is that their own theories are the correct ones

Many flattering remarks have been made about mathematics and the average person has a vague idea that it is one of the few arts or sciences which combine universal general truths on the one hand with complete accuracy of detail on the other. The result is that mathematics is looked on with awe and reverence—and avoided as much as possible. One of the aims of this chapter is to show that mathematics is not so much a science as a language of size and quantity. As such it has all the ambiguities, lack of clarity, and other shortcomings which can be found in any other language. When we are trying to learn an ordinary language we take it for granted there will be a lot of different words which mean the same thing, and often some words which can have two or more entirely different meanings. We expect some of the rules of grammar to be clumsy and contradictory and perhaps even downright silly. The only excuse for these things is that, like Topsy, the language just growed. If we examine mathematics critically we find ground for the contention that it also just growed.

The statement that one plus one is two might seem at first sight a perfect example of a universal and at the same time absolutely accurate truth. Actually it is neither. When we get down to real objects we often find it is impossible to add them together at all. One cow plus one bale of hay might make one contented cow. It might even eventually add up to a couple of gallons of milk but it certainly doesn't add up to two cowbales. Of course it could be argued, providing the cow didn't eat the hay, that one cow plus one bale of hay made two things or two objects. This is quite true. Any one thing plus any other thing will make two things, but we have only achieved a general statement at the cost of becoming so vague as to be practically meaningless. Yet even this vague statement is not universally true. As most of us know, atomic bombs work on the general principle that while small amounts of radioactive material are relatively stable, the build-up of activity in a large amount of such material causes it to disintegrate. Radioactivity builds up in somewhat the same way that heat builds up in a compost heap or a large pile of grass cuttings, only it happens infinitely faster. The point however is that in the case of the radioactive material one quantity plus another equal quantity doesn't make twice the quantity. Even if all the exploded material could be collected it would be found that some of the material had actually become an amount of energy.

From a purely practical point of view examples such as these could perhaps be classed as hair splitting. But the question is not whether mathematics is an excellent practical way of solving many problems but whether there is truth in the notion that mathematics always possesses the universality, precision, and closely reasoned rigorous logic which it is often reputed to possess. In point of fact mathematics has just the same troubles that beset other languages. Take the statement, "Many hands make light work." Like the statement "One plus one is two," at first sight it seems to be an accurate and specific statement which also embodies a fundamental and universal truth. But as soon as we have decided this is so along comes somebody who tells us that "Too many cooks spoil the broth." This is equally true, equally universal, and exactly the reverse of the first statement.

There is another failing which all languages, including that of mathematics, have in common. This is the maddening habit of using one word to mean two or more quite different things. There is the story of a Frenchman who came over to England to improve his pronunciation. He learnt, among other things, that Worcester was pronounced "wooster" but that nobody in the correct circles ever pronounced Gloucester as though it were "glooster." He was getting along very well until one day outside a London theatre he saw a large notice which read, "My Fair Lady—pronounced success." So he packed up and went home.

Mathematics not only does this kind of thing but it goes even further. It has several "dialects," each of them still in everyday use, and in each of these "dialects" the mathematical symbols have their meanings changed just enough to make things thoroughly confusing but not enough to make it easy to realize that there has been any change at all. If the arrangement had been deliberately devised with the idea of making life difficult for the student it could hardly have been organized more effectively. We will examine some of these "dialects," beginning with the simplest and most primitive which for want of a better name we will call "sheep-and-cow mathematics" or "the mathematics of things."

When primitive people first developed the idea of counting it seems pretty certain that the word "one" had a single and definite meaning. It simply meant one object. One sheep plus one sheep made two sheep. They could be large sheep, small sheep, black sheep, or white sheep; but one sheep meant one sheep exactly, neither more nor less. As long as a common name could be found for objects they could be added together. For instance, one sheep plus one cow made two animals. And even though the statement that one thing plus another thing makes two things is so vague as to be practically meaningless the "one" still means exactly what it did before. But while this sheep and cow mathematics is exact and definite it is also very limited. Subtraction, for instance, is confined to the taking of a number of things from an equal or larger number of these things. One can't, in this kind of mathematics, take six sheep from a group of only five sheep. The idea of "minus one sheep," meaning that one person owes another person a sheep, belongs not to the mathematics of things but to the mathematics of human relations—which is another kettle of sheep altogether.

In multiplication, to be precise one should not say that four times five sheep are twenty sheep but that four groups each containing five sheep make a total of twenty sheep. In division a similar outlook applies. The problem is how to divide a group of sheep into a number of groups. Looked at in this light we immediately see that ordinary division will give a realistic answer only when there is no remainder. Also that it can be used only when we want to divide the sheep into groups that are all equal. It is no less reasonable to say that twelve sheep divided into two groups gives one group of seven and one group of five than it is to say that we get two groups of six sheep. So the statement that twelve divided by two equals six is not so universally correct as one might imagine. And as far as getting an odd quarter or half of a sheep left over, the people who work out such answers should be given the job of feeding and rearing these oddments.

As long as people confined themselves to the counting of things their mathematics remained precise and accurate, though limited. But when the question eventually arose of counting the land on which the animals grazed it presented entirely new problems. We cannot count an area of land in the same way that we count a flock of sheep. Man's answer was to devise some sort of measuring stick. Possibly at first the land was described as being so many measuring sticks long and so many measuring sticks wide. Perhaps also about this time people were beginning to engage in barter and were trying to find some accurate way of describing the difference between a large "one sheep" and a small "one sheep."

Some situation such as this must have existed when the mathematics of measurement, as distinct from the mathematics of things, was evolved. Although the development was probably so gradual that it passed almost unnoticed it marked a change as fundamental as the discoveries relating to atomic energy. Man was learning to count and manipulate ideas in the same—or almost the same—way as he had previously counted and manipulated things. As soon as he dropped the conception of a number of measuring sticks lying end to end and began to think in terms of abstract units of length *represented* by a measuring stick he became free from the limitations of the mathematics of things. He had arrived in the realm of abstract ideas.

If we have rows of sheep five abreast and four deep we have twenty sheep. But if we have a piece of ground five measures wide and four measures long we have something quite new—an *area* of twenty *square* measures. A sheep is already a three-dimensional real object, so to talk of a square sheep or a cubic sheep is meaningless. But a foot (the measure, not the thing we walk with) is an abstract one-dimensional concept and when it is multiplied by itself we have a square foot, an idea we can all understand.

Most readers will have heard the story of the woodcutter who boasted that he had used the same axe for thirty years. When pressed for details he admitted it had had seven new handles and three new heads and that the weight of the head and the shape of the handle were different from the original ones. But it was still the same old axe he had bought thirty years ago and he wouldn't use any other. It is something like this which has happened to our numbers and also to the symbols for addition, subtraction, multiplication, and division. Only, to make matters worse, the original "head" and "handle," instead of being discarded, are still being used for sheep and cow mathematics. So we have the situation where two different "axes" are pretending to be one.

We have already seen that one sheep is an exact statement of quantity. But when we come to the mathematics of measurement nobody can produce a length and say it is exactly, absolutely, and precisely one inch or one foot or one anything else. Even if we take a stick and say, "This, by definition, is exactly one foot long," are we going to measure it from the center of the first atom to the center of the last atom, or where? By the time we decide the atoms will have moved anyway. And if we can't hope to set up a precise standard we can hardly hope to produce any other exact lengths when we haven't even anything to work from. So if we are really going to be rigorously scientific we can only say that, as far as we can measure, it seems that a length of one foot as near as we can measure, plus another length of one foot as near as we can measure, appears to approximate to a length of two feet as near as we can measure. Perhaps we can't even go as far as that, for there is the possibility, mentioned by Einstein and others, that length and time are not absolute but relative things. So we could get to the stage of asking, "When is one foot not one foot?"

The ways of using numbers—the processes of addition, subtraction, multiplication, and division—have suffered as much change in meaning as the numbers themselves. Multiplication, for example, in the sheep and cow mathematics, is a kind of repeated addition and this is the definition which is often given. To say that four times two sheep is eight sheep is in fact a short way of finding the total number of sheep when two sheep have another two sheep added to them, and this group of four sheep has another group of two sheep added to them and this group of six sheep has another two sheep added to them.

You can see how much trouble would be saved by multiplication if we were dealing in figures like a thousand times two sheep. When we say that four times two feet is eight feet we are doing almost the same thing, but not quite, for this time the answer is one length, not eight lengths like the eight sheep. But when we take a further step, we find that *four feet times* two feet is eight *square* feet and the definition breaks down. How can one imagine adding two feet to itself *four feet times*? With real objects it can't be done.

However since length is an idea and not a thing we can imagine a length of two feet (such as a piece of string) being dragged sideways for a distance of four feet and so sweeping out an area in the same way that a broom sweeps an area when pushed across the floor. Which is somewhat different from the original idea of repeated addition.

In division, also, the mathematics of measurement is different from the mathematics of things. Since a foot is an abstraction or idea and not an actual thing there is no reason why we cannot imagine it as being divided up into as many bits as we like without its basic quality of length being affected. The tiniest fraction of an inch is still just as much a length as the inch itself. A sheep, on the other hand, if it is divided into small enough pieces, becomes a collection of molecules or atoms which have completely lost all qualities of "sheepness."

The process of subtraction has perhaps had its "head" and "handle" changed more drastically than any of the other mathematical "axes." Previously it was pointed out that in real life one can't have less than no sheep. Minus quantities (often called negative quantities) belong to the world of ideas and therefore it is not surprising to find that they have a logical meaning in the mathematics of measurement which also belongs to the world of ideas.

The meaning of a minus quantity with reference to length is best explained by comparing the subtraction of things such as sheep with the subtraction of ideas such as the idea of length.

If we have six sheep and take five of them away we have one sheep left. If we have a length of six feet and subtract from it a length of five feet we have one foot left. If we look closely we find the two cases aren't the same at all. We don't actually take five things called feet and carry them out of the field and perhaps roast part of one of them for Sunday's dinner! The five feet have disappeared as completely as the Cheshire cat in *Alice in Wonderland.* Not even the grin is left.

But if we stop thinking of the five feet as "things" which are "taken away," and think of the six feet as a distance or length in one direction and the minus five feet as a distance in the opposite direction, we begin to get a picture which does not offend our common sense. We also begin to see that under these conditions it would be quite logical to subtract seven feet from our original six-feet length. We would have an answer of minus one foot which would simply mean one foot in the opposite direction to that of the original six feet.

We can see what happens from the sketch below. We go six feet forward and then seven feet back and finish one foot behind the place we started from.

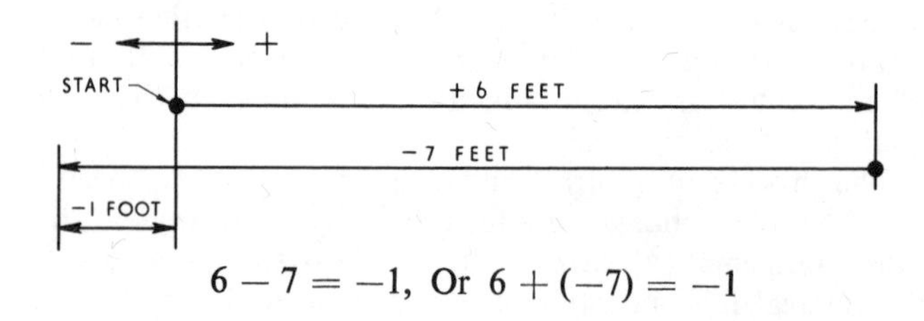

$$6 - 7 = -1, \text{ Or } 6 + (-7) = -1$$

So while subtraction in sheep and cow mathematics merely means the removal of some of the animals to fields and pastures new, subtraction in the mathematics of measurement often means the *addition* of a distance or a force in the opposite direction, rather than the mere removal of an existing force or distance. In fact, when we finish with a minus answer, it can hardly mean anything else.

We have already seen something of the difference between multiplication applied to things and multiplication applied to measurement. The fact that a minus answer is acceptable in the mathematics of measurement makes it possible to make sense of the idea of multiplying with a minus quantity, which according to some mathematicians can't be done. Let us look at the equations below.

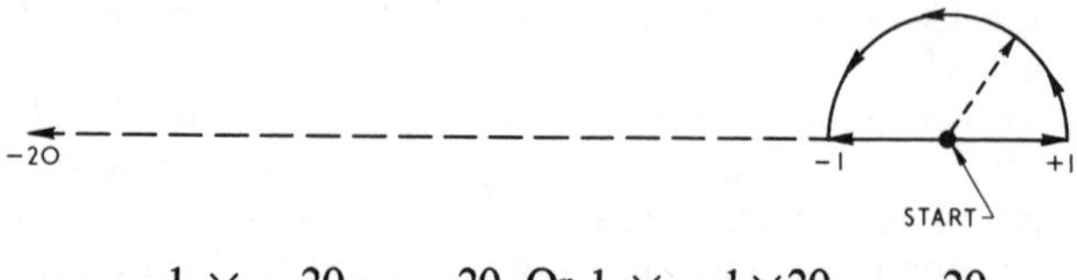

$$1 \times -20 = -20, \text{ Or } 1 \times -1 \times 20 = -20$$

If we multiply a length of one foot by minus twenty we are really doing two things. In the first place we are "swinging" the foot length into the opposite direction and secondly we are "stretching" it to twenty times its original length.

We can now begin to see that both numbers and plus and minus signs, which in the sheep and cow mathematics were a way of describing things, have become, when applied to the mathematics of measurement, also a means of changing the size or direction of these measurements. Numbers and symbols when used in this way are called "operators." In practice it is impossible to draw a hard and fast line between descriptive numbers and operators. One could say, for instance, that 20 is a pure number (if there is such a thing!). On the other hand one could reasonably argue that 20 feet means a unit length (of one foot) operated on by 20 to change it into a length of 20 feet. If we agree that numbers used in this way have a touch of "operator blood" in them we must agree that the minus sign is a veritable half-caste. Under the circumstances it might be interesting to go a bit further and have a look at a "number" which isn't a number at all but a 100 per cent pure-blooded operator. This operator, which usually goes by the cryptic title of "*i*," could be classed as a kind of abominable snowman of mathematics. However, like everything else, it can be understood by common sense and patience.

Briefly, the story is this. Having discovered that the act of multiplying a length by a minus is equivalent to swinging it around through 180 degrees until it is going in the opposite direction, it seems reasonable to ask if it is possible to find some symbol which would swing the length half way round, or in any other direction. Common sense suggests that if the idea of the length being "swung" round is valid, there should be some symbol which would "swing" the length into other directions and which would not clash with the meaning which has already been given to the minus.

In real life we gain our knowledge of an object by looking at it, or touching it, or tasting it, and so on; and also by using it and trying to do something with it. Since mathematical ideas can't be tasted, touched, or seen the only thing left is to try to use them and do something with them. We begin simply by choosing a symbol (any letter or squiggle which is not already being used for something else will do), giving it the meaning and qualities we want, and then we try to use it and see if the results make sense. We will use the title "operator *i*" because that is what it is usually called. Now by definition, when we multiply our length by "*i*" we swing it upward through 90 degrees. The first thing which hits us in the eye is that if we are going to be consistent we must agree that a downward direction, which is opposite to that indicated by plus "*i*," must be indicated by minus "*i*." Again, if multiplying a length by "*i*" swings it through 90 degrees in the direction shown (that is, counterclockwise), then multiplying it by "*i*" again must swing it through a further 90 degrees. And since this will produce exactly the same result as multiplying it by minus 1 in the first place, then it follows that "*i*" multiplied by itself must be equal to minus 1.

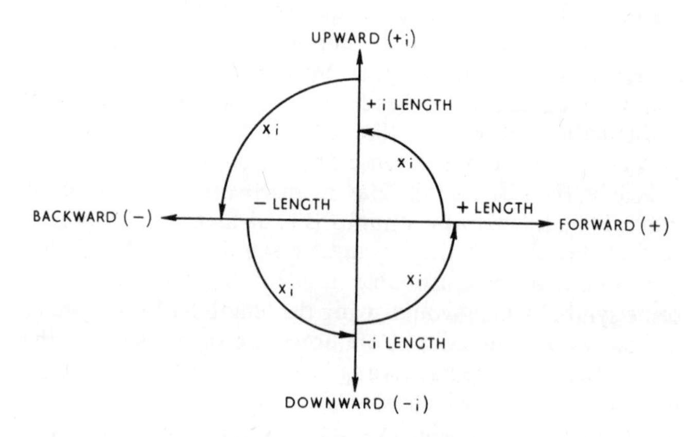

Multiplication by "i" *"swings" the length counterclockwise through 90 degrees, that is, from forward to upward, or from upward to backward, or from backward to downward, or from downward to forward. It does not change the value (or amount) of the length but only the direction in which it is going.*

It is always comforting when working out a new idea to find that something unexpectedly makes sense. If we look again at the downward direction which we decided, to be consistent, would have to be minus i or "$-i$," we see that it is 90 degrees further around than the minus-1 direction is. So we should be able to get there by multiplying minus 1 by "i." This gives us "$-i$" which is what we have already decided the downward direction must be. So there is at least one bit of corroborative evidence to be going on with. If we again multiply this downward direction (which is "$-i$") by "i" we will swing the line back to plus 1, the place we started from. This proves consistent also, for we would be multiplying "$-i$" by "i." This means minus times i by i. Since we already know that i times i is minus 1 we have minus minus 1 and again from our school arithmetic we should remember that this makes plus. So the answer is plus one which is as it should be. The only thing remaining is to find what i is. We already know that i times i equals minus 1.

When any symbol (which includes figures, letters, or any other squiggle we may be using) is multiplied by itself it is said to be squared. The answer is known as the square of the symbol and the symbol is called the square root of the answer.

Thus 3×3, or 3 squared (which is written 3^2) equals 9. And three is the square root of nine. This is often written

$$\sqrt[2]{9} = 3, \text{ or } \sqrt{9} = 3.$$

Similarly i times i or $i \times i$ is the same as i^2 and is equal to -1. So i is equal to $\sqrt{-1}$ just as the 3 is equal to $\sqrt{9}$. Having got to this stage we stop with a jerk, because unlike $\sqrt{9}$ which has an answer, $\sqrt{-1}$ has not. There is no number which when multiplied by itself will give minus 1 except of course $\sqrt{-1}$ which isn't a number in the ordinary sense of the word anyway. So we are left with all lengths in an upward or downward direction having i or $\sqrt{-1}$ permanently attached to them.

If we think it over, however, we find there is nothing really outrageous about this. In fact it is the only thing we could reasonably expect. If i could be worked on mathematically and changed into an ordinary number the distinction between the vertical and horizontal directions would disappear. So i is really nothing more than a mathematical label which makes it impossible for us to get mixed up by adding a length in one direction to a length in another direction (except when they are directly opposite, when the one is minus and the other plus).

After all, to add two lengths which stretched in different directions without taking the direction into account would be as meaningless as adding apples and oranges.

Mathematicians have called $\sqrt{-1}$ (the square root of minus 1) all sorts of unkind names. It is called an imaginary quantity, an unreal quantity, and so on. It is almost like the small boy who said, when he first saw an elephant, "It's not real. I don't believe it." This, more or less, was just what did happen when i was first discovered. Most of the mathematicians of the day, who could think only in terms of sheep-and-cow mathematics, refused to have anything to do with this newfangled monster. It got the same treatment that automobiles and radio and airplanes got when they were first suggested. In all these cases the thing which finally settled the opposition was not "rigorous scientific proof" but the simple fact that people who used these things got practical results which their opponents who wouldn't use them just couldn't get.

It is a sad business that mathematics, which is supposed to be so precise and logical, should hang on to these silly and misleading descriptions, such as "unreal quantity . . . imaginary quantity," and so on. In the first place $\sqrt{-1}$, or i, to use its pet name, is not a quantity at all in the sense in which the word quantity is used in sheep and cow mathematics. But many students, not clearly realizing this, drive themselves silly trying to imagine what the square root of minus one sheep would look like. Even if they manage to bury this piece of confusion safely in their unconscious minds, common sense still asks how can one possibly get a real answer with an imaginary quantity. It is like using an imaginary stick to beat a real dog and then discovering, contrary to all reason, that real hair is flying.

At this stage we had better have a word about brackets. The use of brackets in a formula means that what is included inside the brackets has to be attended to first. If we tell someone to lock up the house and put the cat out we don't expect them to lock the place first and then try to get the cat out through the chimney. Math, however, does not rely on common sense and the order of operations has to be explicitly stated. We have to write in effect "Lock up the house (put the cat out)," or if there were three instructions, "[Lock up the house (put the cat out)] go out shopping."

For instance $6 \times (2+1)$ means that we add the $2+1$ and *then* multiply by 6, giving 18. This is the same as $(6\times2) + (6\times1)$ but is *not* the same as $(6\times2) + 1$ which would be only 13.

Similarly with subtraction $6 - (2+1)$ means we add the $2+1$ first giving $6 - 3 = 3$. This is the same as $6 - 2 - 1$ but is *not* the same as $6 - 2 + 1$, which would be equal to 5. Two or more sets of brackets just means that we attend to what is in the innermost set first and work outward.

But, getting back to i, which is just as "real" as any other mathematical idea, let us see what real, practical, mathematical dogs we can beat with it.

Most people who have cars or live in cities are familiar with maps and particularly with street directories. These nowadays often have numbers and letters along their edges which, in conjunction with an index, assist in finding a particular street.

The map on page 24 is a typical example. If we look up Hunter Street in the general index we find it is at E3. So we look along the bottom edge until we get to 3, and then up the side until we get to E. Where these two levels intersect on the map we find the street we want. This is a quick and handy way of finding any particular place, but it doesn't tell us where that place is with reference to other places on other maps.

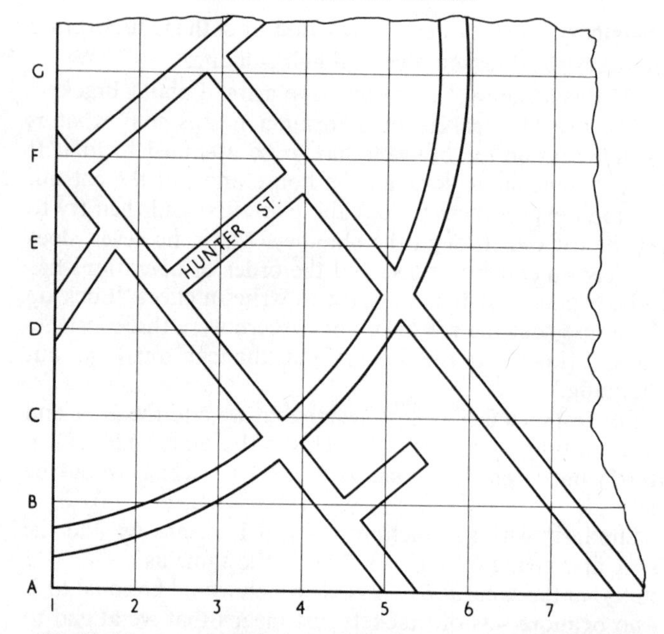

To find how to get there we usually have to trace through several different maps and even then we can only guess at the distance and the direction.

If more people understood the use of the symbol *i* it is possible that the makers of street directories would use this system on the edges of their maps instead of plain numbers and letters. Then instead of each map having a separate numbering arrangement they could all tie in to a system covering the entire area. What is more this system would allow new maps of newly developed districts to be made without any alteration being needed to the numbering system on the other maps. The diagram on the facing page shows the general arrangement of the numbering system.

Also on that page are illustrations of what typical maps of different localities, marked Nos. 57 and 18 in the upper sketch, would be like.

It can be seen that the numbering arrangement on the

two maps is part of the general numbering system and represents miles and decimals of a mile from Fiddlebury Town Hall. It doesn't have to be the town hall. Any other well-known central landmark would do. In some areas, for instance, it might be more convenient to have the numbering

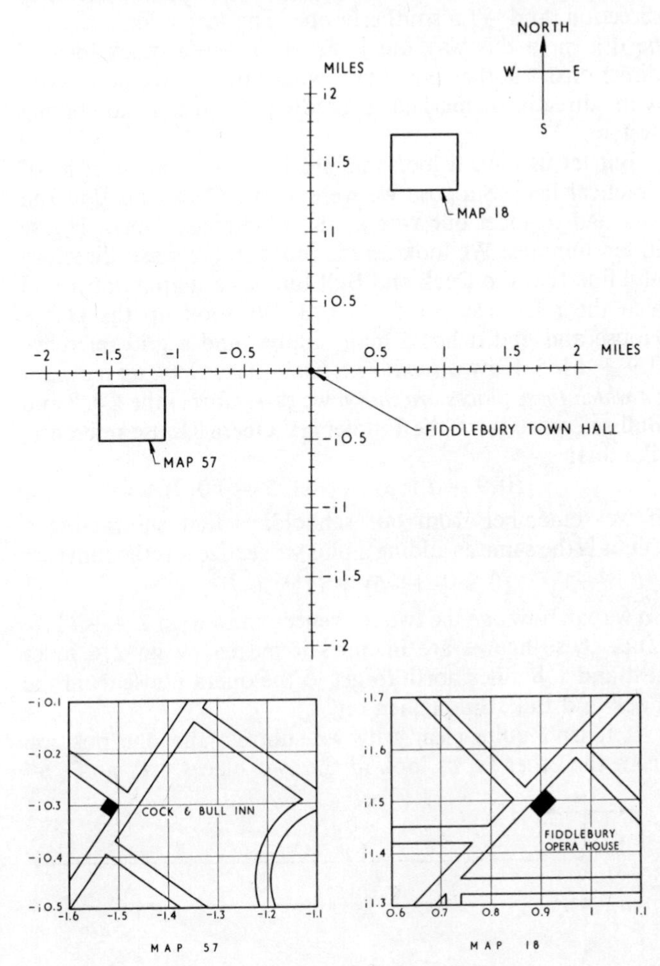

system represent the direction and distance from the central police station. On the general arrangement it will be noticed that the main grid lines run north-south and east-west. This means that i represents north, a positive number represents east, a negative number represents a westerly direction, and $-i$ a southerly one. The maps don't have to be drawn in this way but it helps to give a quick idea of directions. Another possibility would be to have compasses with directions marked according to the i numbering system.

But let us have a look and see how our maps can be of practical help. Suppose we were at the Cock and Bull Inn and had to meet our wife at the Fiddlebury Opera House in ten minutes. We look up the index in the street directory and find that the Cock and Bull Inn has a map number and also the reference $-1.5\ -i\ 0.3$. We look up the Opera House and find it has a map number and a grid reference $0.9 + i\,1.5$. Now *without even bothering to look up the maps on which these places are shown* we can *subtract* the Cock and Bull reference from the Fiddlebury Opera House reference, like this:

$$(0.9 + i\,1.5) - (-1.5 - i\,0.3)$$

If we remember from our schooldays that subtracting a minus is the same as adding a plus we realize it is the same as:

$$(0.9 + i\,1.5) + (1.5 + i\,0.3)$$

So we can now *add* the two references and we get $2.4 + i\,1.8$. Since these figures are in miles it means we go 2.4 miles east and 1.8 miles north to get to the opera house from the Cock and Bull. Simple, isn't it?

If it isn't quite clear why we subtract the one position from the other let us look at the two places "*A*" and "*B*"

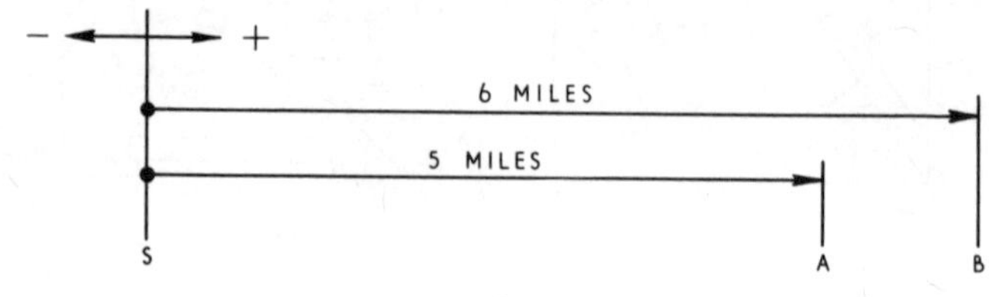

in the sketch below. A is 5 miles from the starting point (marked "S"), and B is six miles from it, in the same direction.

To find the distance from A to B we subtract A (which is 5) from B (which is 6) and get the answer 1. *But* if we want to find the distance from B to A we subtract B (which is 6) from A (which is 5) and get the answer -1. Which simply means that A to B is the direction we have called positive and B to A is in the opposite, or negative direction.

If B was not in the same direction from the starting point as A, but was in the opposite direction, our sketch would look like this.

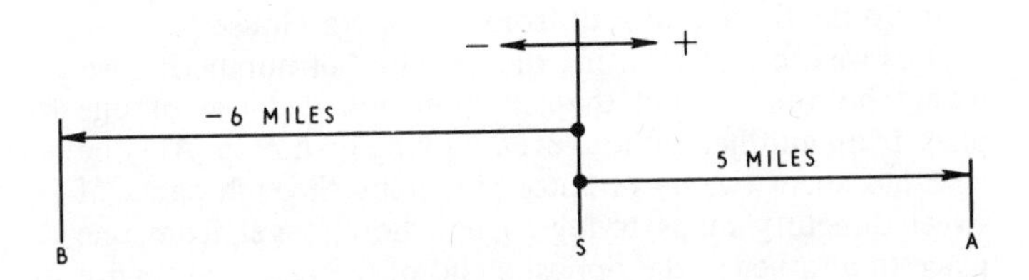

To find the distance from A to B we do exactly the same as before, we subtract A from B. But this time, while A is still 5, B is -6. So taking A from B we have $-6 -5$, which gives -11. Which means, as we can see, that B is now 11 miles away in the opposite direction. If we want to get the distance from B to A we do what we did before and subtract B (which is now -6) from A (which is 5). So taking B from A we get $5- (-6)$. Remembering again that two minuses make a plus we have $5 + 6$, which makes 11. Which means that travelling from B to A we go 11 miles in a positive direction.

This is exactly what we did with the map references in finding the distance from the Cock and Bull to the Opera House.

If we were at the Fiddlebury Opera House and wanted to run over to the Cock and Bull during the intermission we would subtract $0.9 + i\,1.5$ which is the Opera House reference, from $-1.5 - i\,0.3$ which is the reference for the position of the Cock and Bull Inn. Like this:

$$(-1.5 - i\,0.3) - (0.9 + i\,1.5)$$

which gives us:

$$-1.5 - i\,0.3 - 0.9 - i\,1.5$$

which comes to:

$$-2.4 - i\,1.8$$

The figures are of course the same as before because the distance is the same, but now they are both minus which means we are going in the opposite direction. In other words this time we go 2.4 miles *west* and 1.8 miles *south* to get to the Cock and Bull from the Opera House.

Thus we see that by using the *i* method of numbering we would be able to find the direction and distance of one place from another without even looking up a map. Anyone who has spent twenty minutes ploughing through pages of street directory maps trying to find how to get from one place to another at the opposite end of the area will realize what a help it could be. We can also see that our poor old miscalled "imaginary" *i* has a very real meaning and use. It is no more imaginary to talk about a length of *i*5 than it is to talk about five miles north .There is no actual thing or place called north. It is simply, like *i*, a useful and practical label. The fact that *i* is mathematically equal to $\sqrt{-1}$ is sometimes very helpful in calculations but the fact doesn't make *i* a quantity any more than north is a quantity.

The example we have shown of how *i* can be used is a good illustration of how mathematics can make life a lot easier for us if it is applied with common sense and moderation. One could get the same information without mathematics, by finding each place on its respective map and then measuring distances and directions over a number of maps covering the places in between; but it is much more trouble.

Before we leave the street directory maps there is one other matter which should be mentioned. In our calculations we have been finding the relative position of the two places in terms of a distance up or down plus a distance across. If we want to find the actual direct distance in a straight line between the Cock and Bull and the Opera House we have to use the formula we learned at school for getting the hypotenuse (that is the long side) of a right-angled triangle. The method is quite simple, as can be seen from the sketch below. In the triangle shown the long side is marked with the letter *a*.

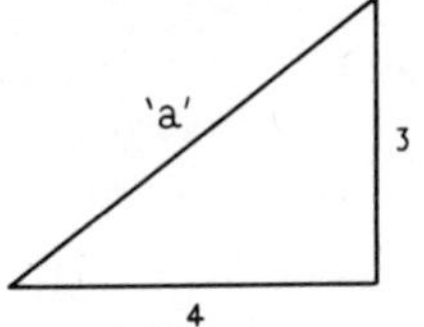

The length of this side *a* is the square root of the sum of the squares of the other two sides, that is $a = \sqrt{3^2+4^2}$.

Since $3^2 = 3 \times 3 = 9$

and $4^2 = 4 \times 4 = 16$

We have $a = \sqrt{9+16} = \sqrt{25} = 5$

In the case of our street maps we have the information that the distance from the Cock and Bull Inn to the Opera House was $2.4 + i\,1.8$. So we have a triangle like this:

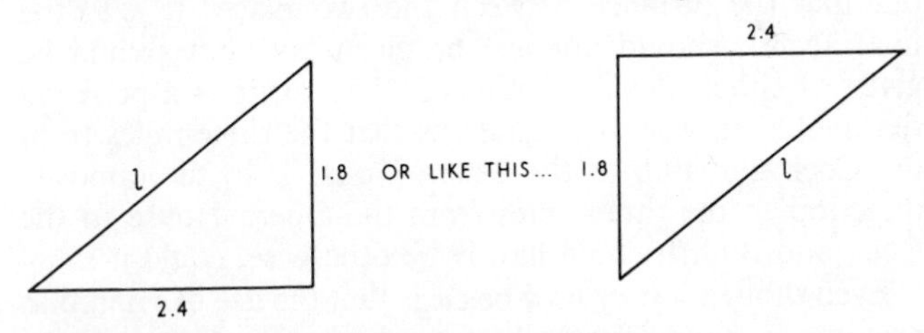

It doesn't matter which shape we use. The long side is exactly the same in both cases and the distance is the sum of the same two sides.

So $2.4^2 = 2.4 \times 2.4 = 5.76$

and $1.8^2 = 1.8 \times 1.8 = 3.24$

So we have $l = \sqrt{5.76 + 3.24} = \sqrt{9.00} = 3$

Common sense tells us that the distance from the Opera House to the Cock and Bull is the same as the distance from the Cock and Bull to the Opera House, although on some occasions it might not seem to be. The distance and direction from the Opera House to the Cock and Bull was $-2.4 - i\,1.8$ so here our triangle has negative sides like this:

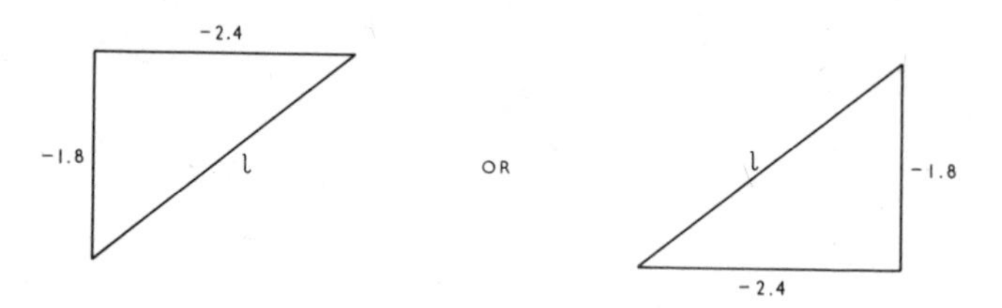

After a little thought we realize it still gives the same result because the square of these sides is minus times minus which gives the same positive answer as does plus times plus.

There remains one other little mathematical point. Since the square of both positive and negative numbers is a positive number, it follows that the square root of a positive number could be either positive or negative. Having worked out that the distance between the two places is $\sqrt{9}$, the final answer should not just be given as 3 but should be given as either plus 3 or minus 3. This again is a perfectly reasonable answer. It reminds us that the three miles from the Cock and Bull to the Opera House is in the opposite direction to the three miles from the Opera House to the Cock and Bull. It could hardly be otherwise, could it?

Even though it may now be clear that the use of *i* notation provides a handy method of working out distances and directions it may seem that we would be just as well off if we used the more customary terms: north, south, east, and west. Instead of saying 3 plus $i2$ miles we could say three miles east plus two miles north. Apart from the fact that it is more cumbersome the trouble would come when we were trying to add or subtract a number of distances. We would have to remember that south is "minus north" and east is "minus west" and vice versa. Also we would always have to make our base lines exactly north-south and east-west or we would soon get into a muddle.

Another advantage of i is that it can be used not only for directions and distances on maps but in exactly the same way to work out problems involving forces of velocities or electrical currents; in fact, for anything which has a direction and a size at the same time. These double-barrelled ideas are called "vectors" and the important thing is that they always have a direction with respect to some basic reference direction. They are a little difficult to visualize because when we think of the size we tend to forget the direction and when we think of the direction we tend to forget about the size. But there is no doubt about their usefulness. The question of how much washing we can put on the clothes line before it breaks, whether a car will skid when going round a corner, what is the safe load which a crane can carry, what size of wire is needed for an electrical system—all sorts of problems can be solved more easily by using the i notation.

In the following sketch we see how easy it is to add a number of vectors together using the i notation. In mathematics it is customary to use a symbol in heavy type to represent the vector and the same symbol in light type to represent the size of the vector. Thus if we had a vector the size of which was eight units, then any line eight units long would represent this size, *provided that it also ran in the correct direction with respect to a reference line*, and would be a vector. In the sketch following the vectors are lines marked **a**, **b**, **c**, etc., and their value in terms of the i notation is shown on the vertical and horizontal dotted lines. We start at the point S and finish at the point F and when we have added all the vectors we find that the sum gives us the distance and direction (called the resultant) from the starting point to the finishing point. We can see the use of this if these vectors represented forces all pulling and pushing in different directions and we wanted to find some single "resultant" force which would represent the total effect of all the other forces. Because these lines represent vectors we have to show a reference line beside the sketch to let us know which direction is *i*, which is minus 1, and so on. In practice this is usually taken for granted.

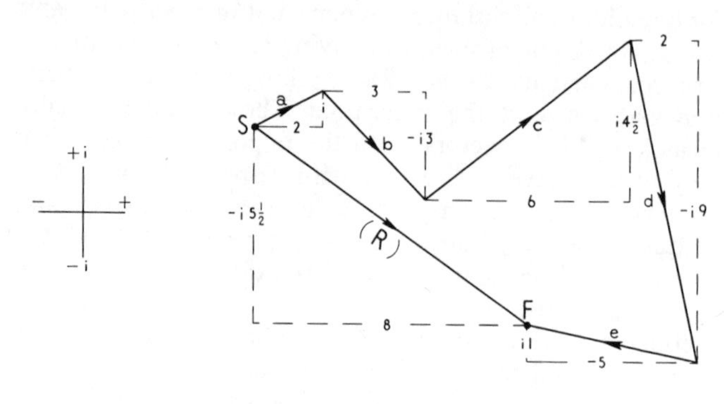

$$\begin{aligned}
\mathbf{a} &= 2 + i \\
\mathbf{b} &= 3 - i\,3 \\
\mathbf{c} &= 6 + i\,4\tfrac{1}{2} \\
\mathbf{d} &= 2 - i\,9 \\
\mathbf{e} &= -5 + i
\end{aligned}$$

SUM $= 8 - i\,5\frac{1}{2}$ WHICH IS THE RESULTANT VECTOR (R) BETWEEN S AND F

The actual resultant force (if these vectors are forces and directions) or the actual resultant distance (if these vectors are distances and directions) is worked out in the usual way, that is,

$$\text{quantity } R = \sqrt{8^2 + 5\tfrac{1}{2}^2}$$

In vector addition we have addition with a new "head" and "handle." If we add a vector which had a value of 1 in one direction to a vector which has a value of 1 in another direction we find that 1 plus 1 can give any answer from zero (if the vectors are in opposite directions and cancel) to 2 (if the vectors are both in the same direction and add).

If they are at an angle of 60 degrees, so that the vectors and the resultant form a triangle with all sides equal, we find that one plus one equals one. If the vectors are at right angles to each other we find that one plus one equals $\sqrt{2}$.

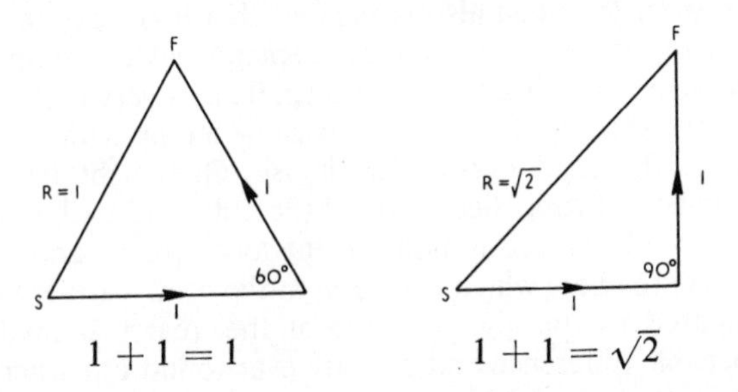

$1 + 1 = 1$ $\qquad$ $1 + 1 = \sqrt{2}$

We must not forget of course that these figures represent only the *sizes* of the vectors and not the vectors themselves, which, as we said before, have direction as well as size.

So much for the addition and subtraction of vectors and the use of the *i* notation. We can do other things with vectors—multiply them for instance—and the results can often have a useful meaning. But it tends to become rather involved so we won't go into it here. Enough has been said to enable us to get a good idea of the basic principles involved in vector mathematics.

Before we finish this section it is worth mentioning another kind of "one" altogether. This crops up in a branch of mathematics known as Boolean algebra which is concerned with the relations between different classes or groups in a closed "universe." In this algebra the total of all classes of things adds up to "one," which represents the whole universe. In this algebra the addition of one plus one is meaningless.

And so we see that the statement "one plus one is two" is neither universally true nor completely accurate. It all depends on what kind of "one" you are talking about.

It may seem confusing at first, but people who can go to the races and find that a gray mare that was considered to be a dark horse has turned out to be a white elephant should have little difficulty in distinguishing between different shades of meaning once they are aware of their existence. If we remember that there *are* these various shades of meaning we can be on the alert for any signs of possible confusion. We must also remember that it is risky to think of ideas such as units of measurement and direction as if they were real things such as sheep. This is very easy to do. Take the story of the man who became fed up with having a hole in the road outside his house. One night he got a strong coil of rope, tied it round the hole, and pulling it out of the road dragged it along to the local quarry and tipped it in. As the hole which had been in the road was very small compared to the gigantic hole of the quarry it made no noticeable difference and nobody ever found out where the hole in the road had gone to.

In case anyone thinks this story is just too silly it could be mentioned that the generally accepted explanation of how transistors work involves the idea of electrons travelling through the material in one direction and holes travelling through it in the other. It's quite a helpful explanation as long as one doesn't take it literally. Don't start asking what happens to the hole which is left by the removal of the hole!

Nevertheless, people do sometimes ask this kind of question and usually get peculiar answers. Sometimes these answers even turn out later to be useful. One could imagine a primitive mathematician playing around with multiplying sheep by sheep and getting square sheep and other peculiarities; and then later generations discovering that these methods, silly when applied to sheep, were excellent when applied to the newly discovered abstract ideas of length and areas and volumes. . . .

Having, it is hoped, made it clear that numbers can have many different shades of meaning we will, in future, unless otherwise stated, use the number language of measurement and for the sake of convenience work with the numbers by themselves. But we must always remember to see that the general ideas are applicable to the particular job in hand before we try to apply them.

It would be fitting to conclude this section by shattering yet another illusion. It was stated earlier that things of different kinds can't logically be added together. Three apples plus two oranges don't make five of anything. One couldn't say there were five appleoranges because neither the apples nor the oranges have changed in any way. They remain three apples plus two oranges. This basic and universal mathematical law has the imposing title of "the theory of dimensions."

In spite of all this the Australian government for one, undaunted by mere academic considerations, takes the weight in hundredweights of all motor vehicles and adds it to the rated horsepower in "horsepowers." It then assumes that in this process of addition both the horsepowers and the hundredweights are magically transformed into identical "power/weight units" and proceeds to levy a very real tax on the answer. One can only say that, while mathematically it is impossible, with tax departments all things are possible.

EXERCISES ON SECTION 1.

1. Make yourself a cup of tea and relax. You are reading this for pleasure and interest, not to pass an exam. Why should you bother with dreary exercises?

2. ONE PLUS ONE IS 10

"They're called lessons," the Gryphon remarked, "because they lessen every day." "The eleventh day must have been a holiday," said Alice. "How did you manage on the twelfth day?" Alice asked eagerly. "That's enough about lessons," said the Gryphon in a very decided tone. . .

In this section we will have a look at the apparently very simple business of counting. Perhaps the earliest method of counting, by primitive people, was to hold up an appropriate number of fingers to indicate the number. Schoolboys still use this method in class to tell each other the time or give any other information containing numbers. If the number happens to be over ten (or five if the person is using one hand) it is necessary to show an open hand several times and finally the remaining amount. When the number of tens became too many to remember easily the obvious thing to do was to keep a record of them, and eventually the counting frame came into being.

At first the recording was probably done just by using heaps of stones as in the sketch below.

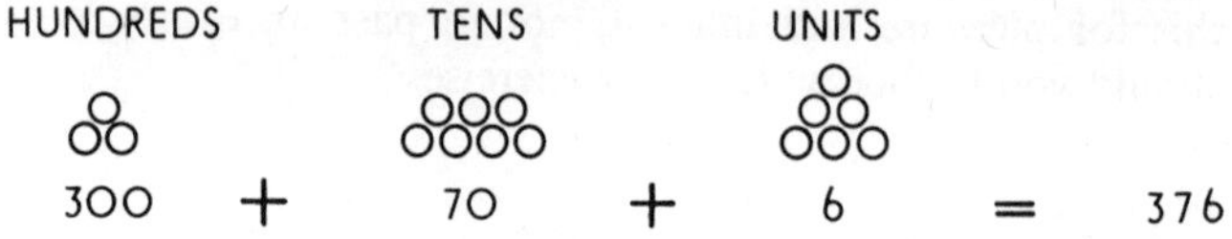

HUNDREDS		TENS		UNITS		
300	+	70	+	6	=	376

For each unit added a stone was added to the right hand pile. When this pile reached ten it was cleared away and one stone was added to the next pile. When this pile reached ten it was cleared away and a stone added to the third, left hand pile. Thus the three piles represented units, tens, and hundreds. There were various numbering systems but this was typical.

For counting, and for simple calculations such as adding,

the primitive people managed well enough. Their troubles began when they tried to write numbers down. All ancient peoples, until the days of the Greeks and Romans, and even later, had a tendency to think of objects—even things like stones and metals—as having a "life" of their own. They also gave the same attributes to ideas, thinking of them not only as things but often as individuals in their own right. This was responsible for two things: firstly, the ancient people had an individual, "personal" name for each number; secondly they thought these numbers had special qualities. We still have a hangover of this kind of thing when we talk of lucky and unlucky numbers. Further, the symbols used for numbers and letters were often the same and a combination of numbers would spell out, or partly spell out, some word. And then all the philosophers would spend years madly trying to find some profound meaning in it. Before we laugh at them too much let us remember how modern "civilized" people will back a horse called "Wide Awake" in the second race simply because a horse called "Nightmare" won the first.

For the ancient people, however, the result was mathematically disastrous. They couldn't do even the simplest calculation on paper. Everything had to be done on a counting frame, and thus only the simplest kinds of mathematics were possible. As far as addition, subtraction, multiplication, and division are concerned a ten-year-old schoolboy can today do calculations which people like Euclid would have considered to be impossible.

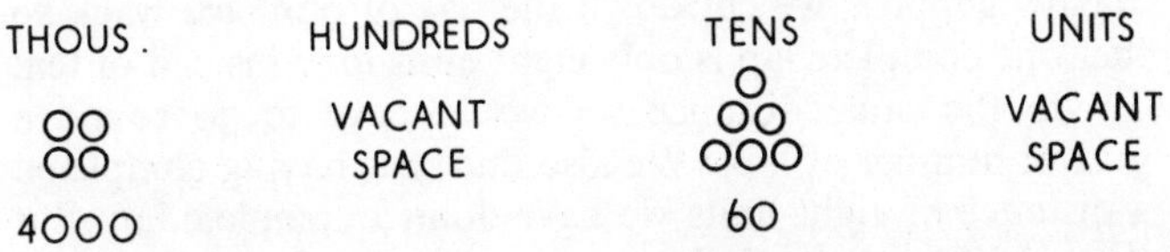

THOUS.	HUNDREDS	TENS	UNITS
4000	VACANT SPACE	60	VACANT SPACE

The great step forward came when the Hindus developed a numbering system which gave on paper a representation of the counting frame or heaps of stones. They invented symbols for one to nine and then, most important of all, invented the nought or zero to indicate a vacant space. The sketch on the preceding page shows this.

Without a zero we could only write this down as 46. It could mean 46, it could mean 406, it could mean 460, or anything else. In this case it is 4060.

The invention of the zero meant that the counting frame could be dispensed with and man could in future work out calculations on paper. Later it was found that the zero not only filled in gaps but was very useful in many other ways.

We probably vaguely remember learning at school that a number, say 257, was made up of two hundreds plus five tens plus seven units. Presented in this way as an isolated piece of information we are tempted to say "So what!" and forget it. However it is worth knowing, because it is almost certainly connected with the fact that we have evolved with ten fingers. If we had only four fingers on each hand instead of five, then 257 would probably have meant the number we now know as 175.

If we think it over we will realize that the fundamental principle behind our method of counting is that of repetition. In our decimal system we count from zero to nine and then begin again. Ten is really zero with a one in front of it to show that we have done one complete lap round a "numerical race track," which in this case is ten units long. When we have completed nine full laps we are up to 90. One more lap and the nine becomes ten with the other zero still in the same position—in other words, we have got to 100. And so it goes on.

Now suppose we cut down the size of our race track so that one complete lap is only eight units long instead of ten. To do the same distance we would have to go round a greater number of laps. We also find that having completed a distance of eight units we have done a complete lap. But the only way we can indicate a complete lap is to chalk up a zero with a one in front of it. So on our smaller numerical race track the counting would go from zero up to seven in the normal way. Then instead of eight we would have to write 10 (one zero, not ten) to indicate that we have com-

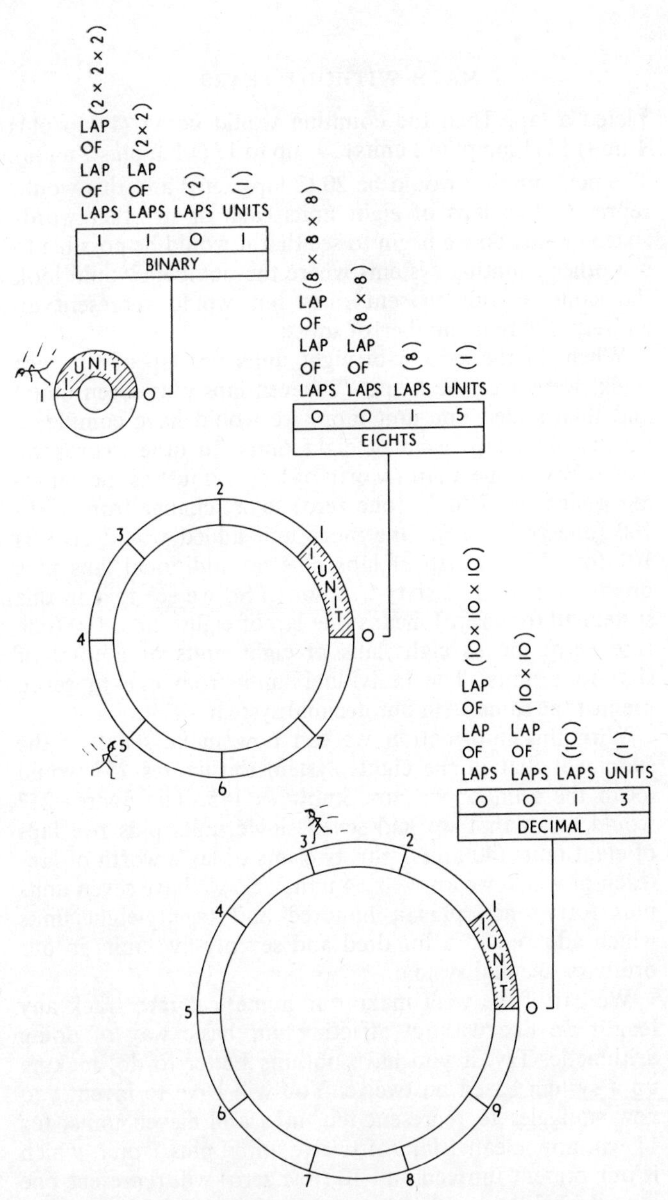

Each man has run 13 units of distance. The smaller the track the more laps he has to go.

pleted a lap. Then the counting would go 11 (1 lap plus 1 unit) 12 (1 lap plus 2 units) . . . up to 17 (1 lap plus 7 units). The next number would be 20 (2 laps zero) and this would represent two laps of eight units each or in other words sixteen units. So we begin to see that it would be possible to use other counting systems where the numbers would look the same as our present ones but would represent an entirely different number of units.

When, in the system of eight units per lap of the race track, we got to the figure 77 (seven laps plus seven units) and then added one unit more we would have completed exactly eight laps each of eight units. In other words we would have done a lap's worth of laps. So just as the counting went from 7 to 10 (one zero) so we change from 77 to 100 (one zero zero). One more unit added would give us 101 (one lap's worth of laps plus no additional laps plus one unit, a total of sixty-five units.) So we see that in this system 10 (one zero) means one lap of eight units, 100 (one zero zero) means eight laps of eight units or a total of sixty-four units. The individual units from one to seven are just the same as in our decimal system.

With this information we can now make sense of the statement that in the eights system the figures 257 would mean the number we now know as 175. The figures 257 would mean that we had seven single units plus five laps of eight units (40 units) plus two lots of lap's worth of laps (each of which we know is 64 units). So we have seven units plus forty units plus a hundred and twenty-eight units which adds up to a hundred and seventy-five units in our ordinary decimal system.

We can if we wish make our numerical race track any length we like without affecting our basic way of doing arithmetic. Try, if you have nothing better to do, making up a system based on twelve. You will have to invent two new squiggles to represent ten units and eleven units, for 11 will now mean 1 lap (of twelve units) plus 1 unit, which is our present thirteen and 10 (one zero) will represent one lap which is now of course twelve units. In this case the figure 100 (one zero zero) would represent twelve laps of twelve units (a lap's worth of laps) which is the normal 144.

The real trouble, however, is that for every different system we would have to learn a completely new set of tables by heart. In the eights system, for instance, twice two would still be four, twice three would still be six, but twice four would be 10 (one zero), twice five would be 12 (1 lap plus two units) and so on. If you have any spare years to waste you can make up such a system, learn all the tables by heart, and astonish your friends at parties by working out sums in your own private mathematical jargon.

There is one system, apart from our decimal system, which is well worth knowing about. This is one which is called the binary system because the lap round the numerical race track is only two units long. You may have noticed that with the eights system we no longer needed symbols for 8 or 9. Now if we make the lap only two units long we don't need symbols for 7, 6, 5, 4, 3, or 2 either. In fact, all we need for any calculation is a 1 and a zero.

This makes life easier in many ways for the people who design computers. The trouble is that in order to make any calculating machine work, we have, in some way, to get it to "understand" differences in the information fed in. If for example we press a key marked "a" on a typewriter and it works exactly the same mechanism as a key marked "b" the typewriter wouldn't be able to "understand" the difference between "a" and "b" and would be useless. An ordinary typewriter has some forty keys which means there must be forty different linkages between the keys and the type bars that hit the paper. If the type bars were operated electrically some distance away from where the keyboard was being operated we would have to have forty separate wires connecting the two. Actually, we could do the same thing with only six wires if we used *combinations.*

The Morse code is used in a similar way. There are only two basic symbols, a dot and a dash, but the combinations of these can cover any letter or figure. The only alternative would be to have about forty dashes, each of slightly different length, and this would be almost impossible to read and would be very liable to error. In the case of the remote-control typewriter we could have one wire and forty different strengths of electric current flowing through it. Or we could have the electric current flowing for forty different lengths of time. But the slightest variation would cause the machine to go haywire.

With about six wires and a coding system we have the advantages of both arrangements. The number of connecting wires is cut down and at the same time we have a perfectly definite arrangement, for the electric current in the wire is either there or not there, it is on or off. When we are doing calculations and working on this kind of mechanical and electrical gadget we find that the binary system exactly suits our needs. We can represent a current flowing as 1 and when the current is not flowing we represent it as 0. A lamp which is lit can be regarded as 1 and a lamp which is out can be regarded as 0. For instance, in the sketch below we have six lamps. Underneath is written what these lamps represent.

In the binary numbering we will see later that this would represent the normal number 41, and it would be possible to get 64 combinations from having various on and off combinations of these six lamps. If we wanted to set up a number in a computer we could have six on-off switches and use the binary numbering system in exactly the same fashion.

In the binary system counting starts from zero to 1 in the normal way. But when we reach two units we have done a complete lap which we write 10 (one zero) as in every other case. Three units then become 11 (one lap plus 1 unit) and four units is our old friend 100 (the lap's worth of laps, two laps of two units).

The tables for binary calculations are simplicity itself. For addition we have to remember that 0 plus 0 equals 0, 0 plus 1 equals 1, and 1 plus 1 equals 0 and carry 1. In other words 1 plus 1 equals 10. Here again we see how mathematical language can get us muddled unless we are careful. The statement 1 plus 1 equals two, and the statement 1 plus 1 equals 10, are both perfectly legitimate ways of saying that one unit plus one unit equals two units.

There is a strong case for believing that our brain does all its basic work in the form of binary calculations by switching electrical impulses off and on. Another department converts this into words, ideas, and so on. Perhaps if we used nothing but the binary system we might be able to release a few million overworked cells from "translating" and put them on to higher duties. Who knows!

As an aid to understanding the workings of these various systems of calculating, a table of comparison is given on page 44, showing the actual number of units (given in the normal decimal system) that some typical symbols would represent in the various systems.

As the binary system is included we can only use zeros and ones for symbols because no others mean anything in the binary system.

One outstanding feature of the table is the difference between the binary system and the others. While 111 in the decimal system means a hundred and eleven units, in the binary system it represents only seven units. From this we might imagine that when using the binary system to represent large quantities we would get a string of figures yards long. Actually it is not as bad as this because the ratio of figures required remains about three to one. One million units in decimals requires seven figures while in binary it requires about twenty. It is often found, especially in computers, that the simplicity of the system gives advantages which outweigh the trouble involved in handling the extra number of figures.

Symbol	1	10	11	100	101	111
'Laps'	1 unit	1 lap	1 lap + 1 unit	1 lap of laps	1 lap of laps + 1 unit	1 lap of laps + 1 lap + 1 unit
Decimal System Numbers	1 unit	Ten units	Ten units + 1 unit = 11 units	Ten lots of ten units = 100 units	Ten lots of ten units + 1 unit = 101 units	Ten lots of ten units + ten units + 1 unit = 111 units
Eights System Numbers	1 unit	Eight units	Eight units + 1 unit = 9 units	Eight lots of eight units = 64 units	Eight lots of eight units + 1 unit = 65 units	Eight lots of eight units + eight units + 1 unit = 73 units
Binary System Numbers	1 unit	Two units	Two units + 1 unit = 3 units	Two lots of two units = 4 units	Two lots of two units + 1 unit = 5 units	Two lots of two units + two units + 1 unit = 7 units

Now let us see how these numbering systems work in practice. We will try multiplying one hundred and forty-three units by twelve units first of all in the usual decimal system, then in the eights system, and finally in the binary system. And to make it clearer we will lay each calculation out in exactly the same way.

In the decimal system we know that one hundred and forty-three is as follows:

One lot of ten-times-ten units	100
plus four lots of ten units	40
plus three units	3
Total	143

Also that twelve is:

one lot of ten units	10
plus two units	2
Total	12

So we have:

```
 143
  12
----
 286
143
----
1716
----
```

or one thousand seven hundred and sixteen units.

In our eights system we see that one hundred and forty three units is:

Two lots of eight-times-eight units	200	(2×64 = 128 units)
plus one lot of eight units	10	(1× 8 = 8 units)
plus seven units	7	(= 7 units)
Total	217	(= 143 units)

Also that twelve units is:

One lot of eight units	10	(1× 8 = 8 units)
plus four units	4	(= 4 units)
Total	14	(= 12 units)

So this time we have as our multiplication:

```
217
 14
---
```

This is where we have to use new tables. Normally four times seven is twenty-eight but in the eights system it is 34 (three laps of eight plus four units), that is 4 and carry 3. Four times one is four, add three makes seven, and four times two is 10 (one lap plus zero). So the first part of our multiplication reads:

```
 217
  14
----
1074
----
```

The second line is 217 as it would normally be. So now we have:

```
 217
  14
----
1074
217
----
```

Now we add it up. 4 plus 0 is 4. 7 plus 7 in this case is 16 (one lap plus six units). 1 plus carry 1 is 2 and 2 plus 1 is 3. So our final sum looks like this:

```
 217
  14
----
1074
217
----
3264
----
```

We then have to find what this means in units.
We have:

	(units in the decimal system.)
4 units	4
plus six lots of eight units	48
plus two lots of eight lots of eight units ..	128
and finally we have three lots of eight lots of eight lots of eight units	1536
Total	1716

Which adds up to the same as before.

Now let us try it in the binary system. In this case it is easier to begin by making a table of how many units the various binary numbers represent or we will be talking of two lots of two lots of two lots . . . so many times that we will become confused and lose count. Computers, by the way, are far better than the "brightest" person in this respect. They never get confused! The values of some of the binary numbers are as follows:

BINARY NUMBER			ACTUAL VALUE IN UNITS
1	-	(no lots of two plus one unit)	1
10	-	2	2
100	-	2×2	4
1000	-	2×2×2	8
10000	-	2×2×2×2	16
100000	-	2×2×2×2×2	32
1000000	-	2×2×2×2×2×2	64
10000000	-	2×2×2×2×2×2×2	128
100000000	-	2×2×2×2×2×2×2×2	256
1000000000	-	2×2×2×2×2×2×2×2×2	512
10000000000	-	2×2×2×2×2×2×2×2×2×2	1024

To get one hundred and forty three units in binary we need:

Taking the numbers from our table:

	10000000	(which is 128 units)
plus	1000	(which is 8 units)
plus	100	(which is 4 units)
plus	10	(which is 2 units)
plus	1	(which is 1 unit)
Total is	10001111	(which is 143 units)

For twelve units we need, again from the table:

	1000	(which is 8 units)
plus	100	(which is 4 units)
Total is	1100	(which is 12 units)

For the multiplication all we need to know is that once one is one and once nought is nought and nought times nought is nought. Simple isn't it? Even a computer can do it.

The multiplication consists of:

```
      10001111
          1100
   -----------
      00000000
     00000000
    10001111
   10001111
   -----------
```

Then we add it up, remembering that one plus one is nought and carry one. Even an educated horse could do the addition!

```
           10001111
               1100
        -----------
           00000000
          00000000
         10001111
        10001111
        -----------
Total:  11010110100
        -----------
```

Having got the answer we can turn again to our table to see what we have got. It breaks up as follows:

```
          10000000000 (which is 1024 units)
plus       1000000000 (which is  512 units)
plus         10000000 (which is  128 units)
plus           100000 (which is   32 units)
plus            10000 (which is   16 units)
plus              100 (which is    4 units)
          -----------           ----
Total is  11010110100 (which is 1716 units)
          -----------           ----
```

This is exactly the same result as we got on the two previous occasions.

What is the value of knowing about these various systems of counting? Firstly it teaches us that there is nothing unique or special about the decimal system. It is just one of many ways of counting. All the same it does give quite a good compromise, as far as human brains are concerned, between a system like the binary, where numbers are hard to remember because there are so many ones and zeros (as mentioned before mechanical memories used by computers don't have this trouble), and a system based on say twenty or thirty, which would need a lot of extra symbols and would have a formidable set of multiplication tables to learn. The second way in which this knowledge can help us is that a thorough understanding of the principles on which our way of counting is based will help us to understand logarithms. Logarithms and the various devices which depend on them can be of tremendous practical use to anyone who needs to do a lot of quick calculations. In some cases they enable us to make gadgets which give us answers without our having to do any calculating at all.

Before we finally finish with the various systems of counting there is one question which some awkward person is bound to raise and that is, where do decimals, or decimal fractions as they are more properly called, fit into these systems? Obviously there must be some way of expressing fractions in all these systems because fractions don't cease to exist just because a different system of counting is used. We will begin by analyzing the ordinary decimal system and then see how the other systems work along similar lines.

In the ordinary decimal system we have already seen that each shift of a number to the left (and putting a zero after it to show this has been done) is equivalent to multiplying it by ten. In the eights system the same thing is equivalent to multiplying it eight times and in the binary system it is equivalent to multiplying it by two. It is obvious that if we do the opposite and shift a number to the *right* this is equivalent to *dividing* it by ten or eight or two as the case may be. There is no reason why we should have to stop when we get to one unit. If we put some boundary mark—a point is the usual one—to show where the whole numbers finish and the fraction begins there is no need to make any other distinction. One shift to the right still is equivalent to dividing by ten (giving us tenths—the next shift divides by ten again (giving us hundredths) and so on. These are written 1.0, 0.1, 0.01, 0.001, etc. Sometimes we just write them as 1, .1, .01, .001, to save trouble. It means exactly the same thing.

Now we have already seen that in the eights system a shift to the left or right means we multiply or divide by eight instead of ten, so just as 10 in the eights system means eight units, so 0.1 in the eights system means $\frac{1}{8}$ instead of 1/10 and 0.01 means $\frac{1}{8} \times \frac{1}{8}$ which is 1/64, and 0.001 means 1/512 and so on. In the binary system the same principle applies but in this case we divide by twos so 0.1 is $\frac{1}{2}$, 0.01 is $\frac{1}{4}$, 0.001 means $\frac{1}{8}$ and so on.

With these facts in mind we are in a position to work out how we can convert an ordinary fraction into decimals, or "eighthmals" or "twothmals" as the case may be. We will probably remember from our schooldays that a fraction such as $\frac{3}{8}$ means three parts out of eight parts and that it is the same thing as 6/16 or 12/32 or even $4\frac{1}{2}$/12. The last is an awkward looking brute but it is quite correct numerically. For convenience, however, we usually multiply this kind of fraction up until we get a whole number on the top and bottom. In this case if we multiply both the top and bottom by two we get the equivalent fraction 9/24, remembering that multiplying both the top and bottom of a fraction by the same number does not alter its value.

We have to apply this knowledge in working out decimals because in the decimal system a fraction can only be so many parts in 10 or in 100 or in 1000 and so on. When we get a fraction such as $\frac{1}{8}$ we find it is one and a bit parts in ten. We can't write this down because it is neither tidy nor accurate so we try if we can get an even figure in hundredths. We find that $\frac{1}{8}$ is $12\frac{1}{2}$/100 which is an improvement. When we try thousandths we find that $\frac{1}{8}$ is exactly 125/1000. This is 0.125. If we want to find the value of $\frac{3}{8}$ we see it is three parts in eight which is three times 0.125 or 0.375. In practice we don't go to the trouble of reasoning it out this way. We have a routine method of working it out, as follows. We take the number in the top line and divide it by the number in the bottom line. Being a fraction the number in the top line is always smaller than the number in the bottom line so the number in the top line has noughts added after it. Every time we add a nought we have to shift the figure in our answer one place more to the right of the unit whole number position. Like this:

8) 3

This won't go, so we add a nought and a point to show the answer has been shifted one to the right:

```
8 ) 30 (0.3
    24
    —
     6
```

Eight into six won't go so we add another nought and remember that the figure in the answer will be the next place to the right. Like this:

```
8 ) 30 (0.37
    24
    —
     60
     56
     —
      4
```

Finally we add another nought and get the full answer 0.375. Like this:

```
8 ) 30 (0.375
    24
    —
     60
     56
     —
      40
      40
      —
      ..
```

If we were to go on dividing we would get 0.37500000... which means 0.375 just as ... 000000000196.000000 ... simply means 196.

Sometimes, however, we don't get a simple answer. The fraction could be one like 1/3 which gives 0.3333333333333 ... until the cows come home. In this case we just go on until the fraction is accurate enough for what we need. It should be clear that the further we go the more accurate our answer will be. In the case above we can see that 0.3 would be 3/10 which is 30/100 and so 0.33 which is 33/100 would be closer. But since 33/100 is the same as 330/1000 we can see that 333/1000 would be closer still to 1/3; and so it goes on.

Returning to our three eights we will see what it is in "eighthmals" that is decimals in the eights system. It might seem obvious that we would start in the same way. Like this:

8) 3

But we can't for the simple reason that there is no such symbol as 8 in our eights system. So we must first convert it. We know that the proper symbol is 10 (one lot of eights plus zero units) So we have:

10) 3

This won't go so we add a nought and a decimal point like this:

```
10 ) 30 (0.3
     30
     —
     ..
```

So we see that the fraction $\frac{3}{8}$ in "eighthmals" is 0.3. Which is obvious really because 0.3 in the eights system is three parts in eight which is what $\frac{3}{8}$ is.

Next let us find out what $\frac{3}{8}$ is in "binamals," or "twothmals," whichever you prefer. Since neither 3 nor 8 are used in the binary system we must begin by converting both of the figures to their binary equivalents. Referring back to the table on page 47 we should be able to work out that 1000 is the equivalent of eight and 11 is the equivalent of three. So this time we have the same song with different words, like this:

1000) 11

Which obviously won't go. So we add a zero and a point in the answer like this:

1000) 110 (0.

But still it won't go. So we add another zero and also this time a zero in the answer to indicate that the first place after the point is empty. Like this:

1000) 1100 (0.0

This time it will go and we work the answer out exactly as we did before. Like this:

```
1000 ) 1100 (0.011
       1000   ——
       ————
        1000
        1000
        ————
        ....
```

Common sense will show us that this answer is correct. As was pointed out before, every time we shift one place to the right in the binary numbering system we divide by two. So starting from 1 which is one unit in any language we see that the first shift to the right, 0.1, is equal to $\frac{1}{2}$, the second shift, which is 0.01, is equal to $\frac{1}{4}$, the third shift 0.001 is equal to $\frac{1}{8}$ and so on. We also notice that our answer, 0.011, is made up of 0.01 plus 0.001, which as we know is $\frac{1}{4}$ plus $\frac{1}{8}$. This of course adds up to the $\frac{3}{8}$, which was the fraction we started with. It is interesting to see that just as we need more whole binary numbers than we do in the decimal system to show whole numbers so we need more "binamals" than we need decimals to show the same fraction.

If you want to do any more experimenting with numbering systems you should now be in a position to work it out for yourself. In every case the basic principles are the same. But the next time that anyone tells you that something is accurate to 0.001 of an inch make sure they are talking in the decimal system, where 0.001 means 1/1000 and not in the binary system where 0.001 would only be $\frac{1}{8}$. Mathematics is such a clear and precise language, isn't it?

3. LOGARITHMS FOR LEISURE

Leisure is time for doing something useful. . . .
Howe

There is a story about a foreman on a building construction job who kept on reprimanding one of the laborers for not getting enough done. Finally the man lost his temper.

"Look, you so-and-so," he bellowed, "I'm working as hard as I can."

"I don't want you to work hard," answered the foreman. "I want you to work fast."

Any sensible person when he has a job to do will take time off at the beginning to find how he can get the most results for the least effort. This is not laziness but intelligence and efficiency. A really lazy person often spends more time and trouble trying to avoid the job than would have been needed to do it. Anyone who is likely to be doing much arithmetic will find that the methods explained in this section can take most of the drudgery out of calculations. The price one has to pay for this increased efficiency is the effort involved in learning and then practicing until these mathematical tools become so familiar that their use is automatic.

In the previous section we saw that various numbering systems could be made with only a few symbols to represent actual numbers and use could be made of their position to indicate how many "laps" round the numerical race track they represented. In the decimal system 10 was one lap of ten units, 100 was one lap of laps or ten lots of ten units, and so on.

As the figures get larger we have to write "ten lots of ten lots of ten lots of ten lots . . . " so often that it wastes a lot of space and time and becomes hopelessly confusing. Writing the figures $10\times10\times10\times$ . . . is a little better but still takes time and space and one can lose count easily. There is however a far easier and better way of writing it which was probably discovered by someone who was either very short of paper or who had a lot of these figures to write and not much time to spare. The idea, like most bright ones, is simplicity itself. Instead of writing down in full the number of figures multiplied together, we just write the figure once with a little note beside it which tells us how often it is multiplied, like this:

Instead of $10 \times 10 \times 10 \times 10 \times 10$ we write 10^5

The small "5" is written high up, about on a level with the top of the main figure, so that there is no danger of confusing 10^5 with 105. It is called a power. One definition for a power is "that which points out or indicates" and in this case the power indicates how many tens have to be multiplied together. There is, of course, no reason why this idea should apply only to tens. In the table of the binary scale in the previous section the final line showed that 10000000000 in binary was equal to $2 \times 2 \times 2 \times 2 \times 2 \times 2 \times 2 \times 2 \times 2 \times 2$. If we count the twos we will find there are ten of them multiplied together. So in the power shorthand we can write 2^{10} which is far easier than the string of twos shown above.

We use the same sort of idea when we are writing in a hurry and put mult^{n} or mult^{y} instead of the full word multiplication or multiply. In the case of figures however the use of a power not only saves time but makes it much easier to read, which is more than can be said for some abbreviations.

To show what an improvement the use of powers can make the table of binary values (page 47) has been rewritten on the next page using powers.

One bright idea often leads to another, and as soon as we write down this table we notice that each time we multiply the actual figures we *add* a unit to the power. If we look a little closer we will see that $2 \times 2 \times 2$, which is 2^3, multiplied by $2 \times 2 \times 2 \times 2$, which is 2^4, will give us $2 \times 2 \times 2 \times 2 \times 2 \times$

2×2, which is 2^7. So if we have $2^3 \times 2^4$ we don't have to write it out in full at all. We just write 2^{3+4} equals 2^7. In the same way 2^7 divided by 2^4 gives us 2^{7-4} equals 2^3.

BINARY NUMBER	POWER		ACTUAL "VALUE" IN UNITS . . .	
1	2^0	(see following pages)	1	UNIT
10	2^1		2	UNITS
100	2^2		4	,,
1000	2^3		8	,,
10000	2^4		16	,,
100000	2^5		32	,,
1000000	2^6		64	,,
10000000	2^7		128	,,
100000000	2^8		256	,,
1000000000	2^9		512	,,
10000000000	2^{10}		1024	,,
and so on . . .				

Any other number or symbol will work the same way. 10^3 which is a thousand, multiplied by 10^2 which is a hundred gives 10^{3+2} equals 10^5 which is a hundred thousand.

Before going on to show what else can be done with powers it is as well to mention one thing we *can't* do. Under no conditions can we add different kinds of powers together. We can't, for instance, say that $3^3 \times 4^3$ is 3^6, or 4^6, or anything else, any more than we can say that 3 apples plus 3 oranges are 6 appleoranges.

Admittedly, as has been mentioned, the Australian motor taxation department has invented "hundredweighthorsepowers" and has torpedoed the theory of dimensions but it hasn't yet tampered with the law for powers.

So it is essential that we remember we can only manipulate powers which all belong to the same number or symbol as the case may be.

Within this limitation however there is practically an open field; and though some of the powers may look a little peculiar they are basically quite reasonable and straightforward.

If, for instance, we have 100 which is 10^2 and divide it by 100 we get 10^{2-2} which equals 10^0. If we have 100 which is 10^2 and divide it by 1000, which is 10^3 we have something even more peculiar looking, namely 10^{2-3} which equals 10^{-1}.

Here again, if we think in terms of the sheep and cow mathematics, where numbers represent things, we can get into a hopeless mental mess. But if we think in terms of the mathematics of measurement, where numbers represent dimensions, we will remember that we can use plus one foot to mean measurement in one direction and minus one foot to represent measurement in the opposite direction.

Using this kind of outlook we can see that if 10^2 means a pair of tens (multiplied together) and 10^1 or 10 means one ten (multiplied together, of course!) then 10^0 means no tens (multiplied together). In other words the ten is cancelled out. And this is what happens, for 10 divided by 10 is $10^{1-1} = \frac{10^1}{10^1} = 10^0 = 1$. Similarly we can see that 10^{-1} is the same as $10^{0-1} = \frac{10^0}{10^1} = \frac{1}{10} = 0.1$. And 10^{-2} is 10^{0-2} which is $\frac{10^0}{10^2} = \frac{1}{100} = 0.01$ and so on.

This kind of thing happens quite often in mathematics. If we use plus 1, plus 2, etc., to mean something we find that minus 1 and minus 2 have a meaning whether we like it or not. In this case 10^{-2} means that ten is "unmultiplied" two times; or, in other words, divided two times. What's more we can't alter matters by pretending a sign isn't there. We can write plain 10 for simplicity but we are really writing $+10^{+1}$ and we must never forget it.

Some mathematicians contend there is "really no such thing as a zero power" and that negative powers "are in themselves quite meaningless," yet they overlook the fact that 10^1, which appears to imply "one solitary ten multiplied together," is just as "meaningless." It is rather like the story of the man who went around with a piece of celery sticking out of his right ear because he argued that no sane person would go round with a piece of celery sticking out of his left ear!

As we saw in Chapter 1, if we accept the idea of a distance being so many feet long instead of thinking of a number of actual measuring sticks laid end to end, and if we accept the idea of length in one direction being positive, we must also accept negative feet, and oddments such as the square root of minus one, representing other directions.

Getting back to the business of multiplying by adding the powers and dividing by subtracting them, we can see that 100×100 is 10^{2+2} and equals 10^4 which is 10,000.

Similarly 10×10 (or $10^1 \times 10^1$) is $10^{1+1} = 10^2 = 100$. When we multiply a number by itself as we did in these two cases we call the answer the "square" of the original number. If we multiply the number by itself and then by itself again we call the answer the "cube." You will notice the connection with "square" feet and "cubic" feet.

If we reverse this process and try to find a number which when multiplied by itself will give the number we started with, then this new number is called the "square root" of the original number. Thus if we start with 10^2 (which is a hundred) and find 10×10 will equal it we can say that 10 is the square root of 100. Similarly 10^2 is the square root of 10^4. And since 10^6 is $10^2 \times 10^2 \times 10^2$ we can say that 10^6 is the cube of 10^2. We can also say that 10^2 is the cube root of 10^6. There is a special sign to indicate roots; it looks like this: $\sqrt{\ }$ for square roots we can put a little "2" in the sign, like this $\sqrt[2]{\ }$ or we can leave it out. So to write in mathematical language that 10 is the square root of 100 we can write $\sqrt{100} = 10$, or $\sqrt[2]{100} = 10$, or $\sqrt[2]{10^2} = 10^1$, or $\sqrt{10^2} = 10^1$ or finally we can write $10^{\frac{2}{2}} = 10$.

Mathematics, which is so often considered to be a shining example of neatness and logic, has more ways of saying the same thing than an exasperated truck driver!

One thing however should be clear from the examples. To get a square, cube, or any other power, we simply multiply the power by 2, 3, 4, and so on. On the other hand if we want to find a square root we divide the power by two, to get the cube root we divide it by three, to get the fourth root we divide it by four, etc.

The fourth power of 10^3 for instance is $10^{3\times 4} = 10^{12}$. And conversely the fourth root of 10^{12} is $10^{\frac{12}{4}}$ which is 10^3. While the sixth root of 10^{12} is $10^{\frac{12}{6}}$, which is equal to 10^2.

In the examples we have shown the power has been a whole number. But this doesn't always happen. Take the cube root of 10^4. It is $10^{\frac{4}{3}}$ which is $10^{1\frac{1}{3}}$ or $10^{1.3333}$. Or the square root of 10, which is $10^{\frac{1}{2}}$ or $10^{0.5}$. These fractional powers make just as much (or as little, whichever you like) sense as the rest of the power system. Don't be disheartened if you notice that $10^{\frac{1}{2}}$ is a power of 10, while $\sqrt{10}$, which is exactly the same thing, is a root of 10. All languages have these helpful little habits!

We can, by ordinary arithmetic, find the square root of 10. It is three and a bit; 3.162, to be accurate. So $10^{\frac{1}{2}}$ equals 3.162 because $10^{\frac{1}{2}} \times 10^{\frac{1}{2}}$ equals $10^{\frac{1}{2}+\frac{1}{2}}$ which is ten; and 3.162×3.162 equals 10 (that is within the accuracy given by three places of decimals).

This makes sense, for if $10^{\frac{1}{2}}$ is to represent any actual number it must be a number which is between 1 (which is 10^0) and ten (which is 10^1), because the $\frac{1}{2}$ power is larger than 0 and less than 1.

Similarly we would expect to find that $10^{1.3333}$ is a number larger than ten (10^1) and less than 100 (10^2). This turns out to be so, for the cube root of 10^4 (ten thousand) which is $10^{\frac{4}{3}}$ or $10^{1.333}$ is about 21.54. So $10^{1.333}$ represents the number 21.54.

Let us make a list of the various powers we have been talking about and their values. Like this:

$$10^{-2} \quad = \quad \frac{1}{10^2} \quad = \quad 0.01$$

$$10^{-1} \quad = \quad \frac{1}{10} \quad = \quad 0.1$$

$$10^0 = \frac{10}{10} = 1.0$$

$$10^{\frac{1}{2}} = \sqrt{10} = 3.162\ldots$$

$$10^1 = 10 = 10$$

$$10^{1.333} \quad \sqrt[3]{10^4} \quad 21.54\ldots$$

$$10^2 = 10 \times 10 = 100$$

Now let us fill in some figures between 10^0 (1) and 10^1 (10) It can be worked out that:

$10^0 = 1$ $\qquad$ $10^{0.7781} = 6$

$10^{0.3010} = 2$ $\qquad$ $10^{0.8451} = 7$

$10^{0.4771} = 3$ $\qquad$ $10^{0.9031} = 8$

$10^{0.6021} = 4$ $\qquad$ $10^{0.9542} = 9$

$10^{0.6990} = 5$ $\qquad$ $10^{1.0} = 10$

These figures were taken from a table of logarithms. This sounds something like the statement of the small boy who, when asked where sugar came from, said, "From the local grocer." Actually it is not as silly as it sounds as long as we realize it is only half the answer. Anyone who wants to know how these figures are worked out can look up any standard textbook on the subject.

For those who want to be satisfied there is no jiggery-pokery or sleight of hand going on, it can be pointed out that there is a way of getting these figures by simple arithmetic, common sense, and an awful lot of patience.

Here is one way to go about it, taking for example 2, which is supposed to be $10^{0.3010}$. This is approximately equal to $10^{\frac{30}{100}}$.

So we find by trial and error a number (it will be about 1.0023) which when multiplied by itself a hundred times will equal 10. Having got this—it may only take a month or so—multiply it by itself thirty times and the answer should be very nearly two. What we have done is simply to get the hundredth root of ten, which is

$$\sqrt[100]{10} \text{ or } 10^{\frac{1}{100}}$$

and multiply it to the 30th power which makes

$$10^{\frac{30}{100}} \text{ or } 10^{0.30}$$

But without all this trouble our common sense will provide a good check. 2^3 is $2\times2\times2$ which is eight. So two and a bit would be the cube root of ten; in other words, $10^{0.333}$ . . .. So it seems reasonable that $10^{0.3010}$ is 2.

Again, the square root of ten, which is $10^{\frac{1}{2}}$ or $10^{0.5}$, is about 3.162 . . . So we would expect, if $10^{0.5}$ is 3.162, that $10^{0.4771}$ is likely to be equal to 3.

If we check the other figures in the same way we will find they are what we would expect them to be.

This sort of "proof" is calculated to make any orthodox mathematician's hair stand on end. It might not help much if we wanted to calculate a complete table of logarithms. But who wants to? If one drives a car it helps to have a good practical knowledge of how and why it works. But it is a waste of time knowing every dimension of every part and the manufacturing tolerances and the chemical composition of every material. It is the same with mathematics. We may want to use it, but we certainly don't want to manufacture it from scratch.

Assuming, however, that the table on page 61 is correct we will see how it can be used. You will remember that we can multiply 100 (10^2) by 1000 (10^3) by adding the powers thus 10^{3+2} equals 10^5. We can do exactly the same thing if we want to multiply, say, four by two. Four is $10^{0.6021}$ and two is $10^{0.3010}$. So four times two is $10^{0.6021+0.3010}$, which equals $10^{0.9031}$ which from the table we see is eight.

Let us try again this time multiplying five by four. Five is $10^{0.6990}$ and four $10^{0.6021}$. So five times four is $10^{0.6990+0.6021}$, which gives $10^{1.3011}$. Now our table doesn't give any number for $10^{1.3011}$, but let us write it another way, like this, $10^{1+0.3011}$. This means ten times $10^{0.3011}$ which is near enough to 2. So 4×5 is two times ten, which is 20. In the same way fifty times forty would be $10^{1.6990+1.6021}$ equals $10^{3.3011}$, which is $10^{3+0.3011}$, which is two times a thousand, which is 2,000.

We can do this because ten is the base for our powers as well as the "lap" for our counting system. We don't have to use ten for either of these, but it makes calculations a lot easier if we do.

You may have noticed that when we added the powers for four times five we get 1.3011 whereas the table shows that for 2 it should have been 0.3010. Actually the figure, correct to five places, is 0.30103. Tables correct to seven figures and to ten figures and even more can be obtained but for most purposes the four-figure tables are quite good enough. In practice it means that a correct answer could be either 4672 or 4673, or if it was a decimal 0.4672 or 0.4673. Or, if it was in the forty thousands, it could be anywhere between 46720 and 46730 or even as low as 46715 and as high as 46735. But what is a mere twenty out of forty thousand between friends?

Division works the same way as multiplication except that we subtract one power from the other.

For instance, 8, which is $10^{0.9031}$, divided by two, which is $10^{0.3010}$, is $10^{0.9031-0.3010}$, which is $10^{0.6021}$, which is 4.

When we get to negative powers, however, we have to be careful. In ordinary subtraction of one quantity from another what we really do is find the *difference* between them and make it positive or negative according to whether the first or second quantity is greater.

Like this: $0.7781-0.4771 = 0.3010$.

But if it is $0.4771-0.7781$ we still get the same amount of difference but put a minus sign in front of it thus -0.3010.

Let us try this with powers and see what happens.

The first one is easy. $10^{0.7781}$ which is 6, divided by $10^{0.4771}$ which is 3, gives $10^{0.7781-0.4771}$ equals $10^{0.3010}$ which is 2.

If we divided 3 by 6 the powers would be the same and the answer would be $10^{-0.3010}$. It was pointed out on page 58 that a minus power meant a fraction so $10^{-0.3010}$ simply means $\frac{1}{10^{0.3010}}$, which is $\frac{1}{2}$, since $10^{0.3010}$ is 2.

Now this is a perfectly correct answer but it has the disadvantage that it is not a decimal fraction. It can be a nuisance trying to add up say 3.8472 plus $\frac{1}{48.37} + \frac{1}{563}$ and so on.

There is however a neat way in which we can get the answer in decimals without much trouble. Let us go back to 3 divided by 6 which is $10^{0.4771-0.7781}$. If at this stage we first multiply by ten and then divide by ten it can't affect the answer. This of course only means adding and subtracting 1 from the power. So we can write it like this $10^{0.4771-0.7781}$ equals $10^{+1+0.4771-0.7781-1}$. Add the 1 to the 0.4771 and then subtract 0.7781 and we get $10^{1.4771-0.7781-1}$. This comes to $10^{0.6990-1}$. Now $10^{0.6990}$ equals 5, multiplied by 10^{-1} which is $\frac{1}{10}$ so we get $\frac{5}{10}$ or 0.5 which is our $\frac{1}{2}$ in decimals.

In the last example it may have seemed we are going to a lot of trouble to get 0.5 instead of $\frac{1}{2}$ which is known to be 0.5 anyway. Let us take another example. Divide 9 by 10. We know the answer is $\frac{9}{10}$ or 0.9 but let us use powers. From our table on page 61, nine is $10^{.9542}$ while ten is 10^{1}. So we have $10^{0.9542-1}$ equals $10^{-0.0458}$. From a book of tables we find that $10^{0.0458}$ is 1.1112. So our answer $10^{-0.0458}$ is $\frac{1}{1.1112}$ which is an extremely unpleasant-looking fraction. One would hardly recognize it as being the same as $\frac{9}{10}$ but it is!

Doing it the other way we have $10^{1.9542-1-1}$ equals $10^{0.9542-1}$ which equals $\frac{9}{10}$ which is 0.9.

In all the examples so far we have written a ten with a power like this: 3 equals $10^{0.4771}$.

The 0.4771 is what we are concerned with and the 10 can be dropped for our practical calculations.

So we adopt a jargon and say that

0.4771 is the logarithm of 3, and that
3 is the antilogarithm of 0.4771

These are to the base 10. There are other tables of logarithms to other bases but they are not often used except for special jobs.

What we have done so far in this section may seem interesting but not very useful. The following examples are intended to show how useful logarithms can be.

First however a note about how logarithms are usually written down. For any number between 1 and 10 we write it straight from the table with a decimal point in front. Thus:

log 3 equals 0.4771

If the number is between 10 and 100, say 30, it will be log 3 plus log 10 (which is 1), thus:

log 30 equals 1.4771

If the number is between 100 and 1000, say 300, it will be log 3 plus log 100 (which is 2), thus:

log 300 equals 2.4771

We have to be careful when working with the logarithms of numbers less than one. Although the principle is exactly the same, the fact that we run into minus quantities tends to frighten beginners.

Just as thirty (3×10) is log 3 plus log 10 so we find that 0.3 ($3 \div 10$) is log 3 minus log 10.

As explained on page 64 there are two ways in which we can write this down. We can get log 3 (which is 0.4771) and subtract log 10 (which is 1) from it. Like this:

$0.4771 - 1$ equals -0.5229

Now since the antilogarithm of 0.5229 is 3.3333, the antilogarithm of -0.5229 will be $\frac{1}{3.333}$ and we finish with a fraction.

It is easier to use the second way which is simply to leave the two parts separate as $+0.4771-1$. This is equivalent to leaving it as $3 \times \frac{1}{10}$ (or $3 \div 10$, whichever you prefer).

Another advantage of this way is that it ties up with what we already do with positive numbers. Thirty, as we mentioned above, is log 3 plus log 10, which is $+0.4771+1$. This of course is the same as $+1.4771$. *But* what we must get clear and always remember is that $+0.4771-1$ is *not* equal to -1.4771. If we draw a diagram like those in Chapter 1 we will see the difference.

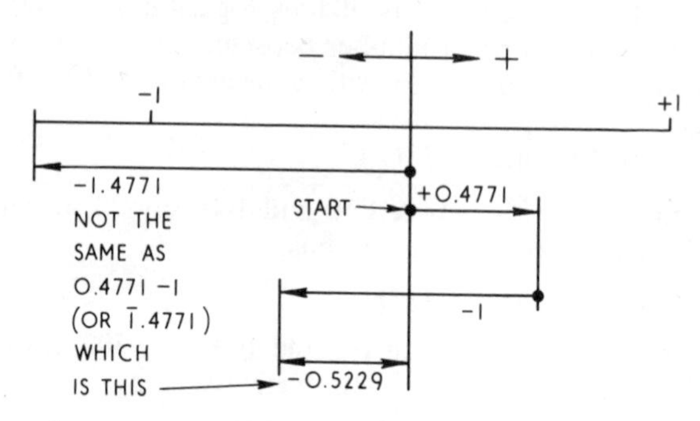

So unless we are going to get our result in the form of an awkward fraction instead of decimals we must be content to leave it as $+0.4771-1$, or as $-1+0.4771$ which is exactly the same thing.

The fact that $+1+0.4771$ can be written simply as 1.4771 led some enterprising people to wonder if a neater way of writing $-1+0.4771$ could be found, and the following method was eventually adopted. Since the minus sign couldn't be put in front of the composite number it was put *over* the negative part of it, like this: $\bar{1}.4771$.

This minus or "bar" logarithm is just a simple way of writing down the composite number and if we remember this we should not have any trouble working with negative logarithms. The table below shows how conveniently this method fits into the general pattern.

Log 3000 $(3\times1000) = +0.4771+3 = 3.4771$
Log 300 $(3\times100) = +0.4771+2 = 2.4771$
Log 30 $(3\times10) = +0.4771+1 = 1.4771$
Log 3 $(3\times1) = +0.4771+0 = 0.4771$
Log 0.3 $(3\times\frac{1}{10}) = +0.4771-\bar{1} = 1.4771$
Log 0.03 $(3\times\frac{1}{100}) = +0.4771-2 = \bar{2}.4771$ and so on.

When it comes to using logarithms in calculations we will find that, once the basic ideas are understood, the rules are simple. If we want to multiply numbers we add their logarithms. If we want to divide one number by another we subtract the logarithm of the one from the logarithm of the other. If we want to square or cube a number we multiply its logarithm by 2 or 3. If we want to get a square root or a cube root we divide the logarithm by 2 or 3.

If we want to multiply the following numbers together

$$1.784 \times 89.4 \times 309.3$$

Log 1.784 equals	0.2514
Log 89.4 equals	1.9513
Log 309.3 equals	2.4904
TOTAL LOG	4.6931

The antilogarithm of 0.6931 is 4.933 and the 4 means we multiply it by 10^4 (10,000). So the answer is 49330.

As explained before with four-figure logarithms the answer could be between about 49320 and 49340. If we want the last two figures to be right we could use tables correct to six or seven figures.

Division is simply a matter of subtraction. The following example, where a small number is divided by a large one, illustrates what has already been said about using a negative whole number and a decimal part which is always kept positive (the whole number, by the way, is sometimes called the "characteristic" part of the logarithm, and the positive decimal part is sometimes called the "mantissa").

For example, let us divide 31.8 by 473.

Log 31.8 equals	1.5024
Log 473 equals	2.6749
Subtracting gives	$\bar{2}.8275$

which is the antilog of 0.8275 multiplied by $\frac{1}{100}$, which is $\frac{6.722}{100}$. So the answer is 0.06722.

If you are still worried about the "bar," then try doing the sum this way:

$$\text{Log } 31.8\ (\ 10\times3.18) = +1+0.5024$$
$$\text{Log } 473\ \ (100\times4.73) = +2+0.6749$$

Since the second part—the mantissa—must always finish positive to prevent our getting a fractional answer we rewrite the sum like this, borrowing 1 from the characteristic:

$$+1+0.5024 = +\not{1}\ -\not{1}\ \text{(borrowed)}\ +1.5024$$
$$+2+0.6749 = +2, \qquad\qquad +0.6749$$

Subtracting we get

$$-2 \qquad\qquad +0.8275$$

which as previously explained is written $\overline{2}.8275$

If we want to get the cube root of a number we simply divide the logarithm by 3. For example, what is the size of a cube-shaped box which would enclose a volume of 85 cubic feet?

Log 85 equals 1.9294
Divide by 3 equals 0.6431

Antilog of 0.6431 equals 4.396 feet, which is the answer.

We have to be a little more wary when dealing with a fractional number. For instance, take the cube root of 0.85.

Log of 0.85 is $\overline{1}.9294$

To divide by 3 here we must first rewrite the log like this

$\overline{3}$ plus 2.9294

Now we can divide by 3. This gives $\overline{1}.9765$, which is 0.9473.

What we have really done is to rewrite 0.85 (already written as $8.5 \times \frac{1}{10}$) in logarithmic equivalent as $850 \times \frac{1}{1000}$ This is exactly the same thing but it enables us to get separately the cube root of 850 (which is 9.473) and the cube root of $\frac{1}{1000}$ (which is $\frac{1}{10}$) and then combine them ($9.473\times\frac{1}{10}$) to get our answer, 0.9473 in decimal form.

If instead we divided the -1 and the $+0.9294$ by 3 we would have $-0.3333+0.3098$, which equals -0.0235. The antilogarithm of this is $\frac{1}{1.1055}$, which is 0.9473 in an unmanageable fractional form instead of a convenient decimal one.

So we see that the use of logarithms converts multiplication and division into the much simpler business of addition and subtraction. Working out roots and powers of any number becomes very easy, instead of being, in the case of most roots, a lengthy trial and error process.

As with anything else, one only becomes skilled at using logarithms after a certain amount of practice. If you want to learn to use them keep practising and don't get depressed if someone else seems to learn faster than you.

Some people are naturally good at skating, baseball, or tennis. Others are naturally quick at learning to use mathematics. In none of these cases does it automatically mean that the person concerned has a high degree of general intelligence and common sense. Sometimes in fact the opposite can be nearer the mark.

The next section will show how logarithms can be used for the construction of gadgets and charts which take all the calculating out of calculations.

With these all you have to do is to lay a straight edge across a chart or slide one graduated scale past another and, presto, you have the answer.

But before we leave logarithms and powers there is one interesting sidelight. You will remember our "imaginary" friend "*i*" in Chapter 1. If plus 1 indicated a direction in one way and -1 indicated the opposite direction we discovered that *i* indicated a direction at right angles. Like this:

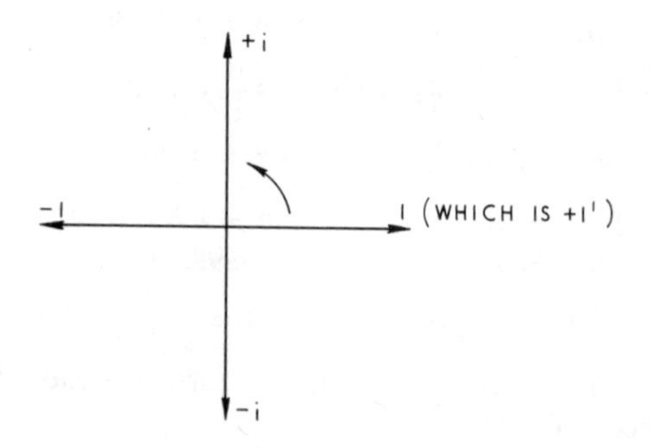

We also discovered that i^2 was equal to -1. In this chapter we have found that numbers can have fractional powers; for example, the square root of 10 can be expressed as $10^{\frac{1}{2}}$. Why shouldn't the same thing apply to i and if so what would an expression such as $i^{\frac{1}{2}}$ mean if anything? Firstly it would mean the square root of i, which is the square root of the square root of -1, or $\sqrt[4]{-1}$ or $-1^{\frac{1}{4}}$. All of which leave us about as wise as before. If, however, we think of i as being an operator swinging a line around, we begin to see that just as $i \times i$ equals -1, so $i^{\frac{1}{2}} \times i^{\frac{1}{2}}$ equals i. So if i represents an operation of swinging the line through 90° then $i^{\frac{1}{2}} \times i^{\frac{1}{2}}$ represents the swinging of the line through 90° in two equal operations. In other words, $i^{\frac{1}{2}}$ represents swinging the line through an angle of 45°.

This makes sense too because i^0, like anything else to the zero power, is plus 1 and i^1 represents a counterclockwise swing of 90, so $i^{\frac{1}{2}}$ being half-way between i^0 and i^1 ought to represent a swing of 45°. In the same way we can see that any fractional power of i will represent a certain direction.

Let us look at the sketch on the facing page.

According to what we found in chapter 1 the *distance and direction* from A to B should be 1 plus i. If what we have said above is true, since it is a 45° triangle, the direction from A to B should be $i^{\frac{1}{2}}$ and the distance $\sqrt{2}$ (that is $\sqrt{1^2+1^2}$) so the *distance and direction* should also be equal to $\sqrt{2} \times i^{\frac{1}{2}}$.

Therefore $\sqrt{2} \times i^{\frac{1}{2}}$ should equal 1 plus i.
If we square both sides of this equation we have

$$\sqrt{2} \times i^{\frac{1}{2}} \times \sqrt{2} \times i^{\frac{1}{2}} = (1+i)(1+i)$$

This gives $2i = 1 \times 2i + i^2$.
Since $i^2 = -1$

We get $2i = \cancel{1} + 2i - \cancel{1}$

$\therefore\ 2i = 2i$

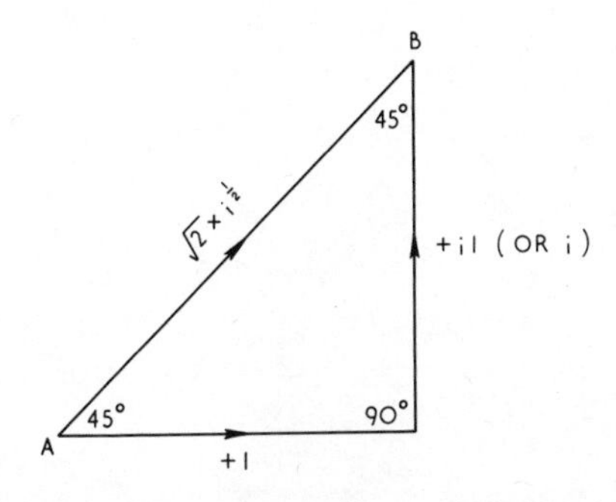

This is the sort of proof that the mathematician delights in. He fiddles with both sides of the equation in all kinds of ingenious ways until he reduces both sides to the same quantity. It's perfectly legitimate as long as he does exactly the same thing to both sides. If he multiplies one side he must multiply the other side by the same thing. If he squares one side he must square the other.

These maddeningly unanswerable proofs are as indisputable as the rabbit which the magician pulls out of his hat. The magician starts with an obviously empty hat, goes through a series of complicated movements, and produces a perfectly genuine rabbit. The mathematician starts with two quantities which haven't the slightest resemblance to each other, goes through a complicated business of multiplying them, converting them, squaring them, and finally reduces them to two obviously identical quantities, which proves the original quantities must also have been identical.

Nothing is more calculated to raise a sense of frustration and inferiority in the bosom of the genuine hater of mathematics, and so we will do our best to avoid allowing any more of such proofs to creep into this book. But this one was irresistible!

4. ABACUS AUTOMATION

I like work; it fascinates me.
I can sit and look at it for hours.

Jerome K. Jerome

Philosophers and others have spent a great amount of time trying to define exactly what qualities distinguish man from what we are pleased to call the lower animals. The problem is more difficult than we might imagine. But if we had to choose one distinguishing characteristic of man which is not—as far as we can see—possessed to any extent by other creatures, we could well choose the faculty of conscious imagination.

It is true that animals at play show some of the signs of what in humans would be called imagination. A kitten will play with a cork or a ball of wool exactly as though it were a mouse. And we have no way of telling just to what extent the kitten "knows" that the cork or the ball of wool is a substitute for the real thing.

Without the possession of imagination man would never have been able to learn to speak or to write, much less to invent and use mathematics. For in all these things imagination is essential if we are to recognize the various sounds and squiggles as being symbols for other things; especially because, more often than not, there is no physical resemblance between the symbol and what it represents.

At the beginning of Chapter 2 we gave an indication of how primitive man might have begun his first attempts at counting by sorting pebbles into groups and imagining that each pebble in the first group represented one sheep, or one cow, etc., and each pebble in the second group was the equivalent of ten pebbles in the first group, that is, it represented ten sheep, or cows, or other things being counted.

If we think it over, we realize this is quite a feat of imagination, for not only does one object represent another object but when in the second group it represents a number of objects.

There are quite a few people who, while they would regard the use of heaps of stones for counting objects as childishly simple, become confused when the word "algebra" is mentioned, and say they never could understand it. And yet, on the whole, it requires less imaginative effort to write down some letter such as "*a*" and say that it represents any number (or that it represents a number which we are trying to find) than it does to pick up a pebble and say, "This represents one (or ten, or a hundred) sheep."

Perhaps the reason why so many people give algebra away is that—like most branches of mathematics—it has an imposing appearance and a conglomeration of muddle and inconsistency underneath. For example, let us take two numbers, 6 and 3. Now if we want to multiply them together we write 6×3 (which is 18). If we write down 63 it means something entirely different (6 tens plus 3 units). But in algebra, if a and b represent two numbers and we want to multiply them, the usual way is to write ab, not $a \times b$. Again 106 is one hundred and six, but $10a$ is ten times a or $10 \times a$. In division on the other hand we usually say $a \div b$ or $\frac{a}{b}$ in the same way that we do with any actual numbers which a and b may represent.

We can no more blame mathematics for having these confusions and contradictions than we can blame a dog for having fleas. In both cases it is a sign of a neglected upbringing, but it does make life difficult.

We will try to ease the muddle as much as possible, in future examples, by always writing $10 \times a \times b$ instead of $10ab$ and so on.

We have already seen a number of examples where, if we take a little trouble to learn a mathematical method, we can make difficult problems into simple ones and in the long run save time and trouble.

Algebra has the same usefulness. Suppose we have a room 7 feet wide by 10 feet long. We know the area is $7 \times 10 = 70$ square feet. If the room is 8 feet wide by 12 feet long we have an area $8 \times 12 = 96$ square feet.

We can go on giving specific examples forever without getting any further. What we need is some general rule which can be applied to all possible cases whatever the length and width may be. We could get this by putting in the following words: "A rectangular shaped room with any width at all and any length at all would have an area which would be the product of the width and the length. In the cases under consideration the width and length would both have to be in feet and/or parts of a foot and the area would then be in square feet and/or parts of a square foot." This takes 67 words and is rather hard to follow. Using algebra we can say, "A rectangular shaped room '*a*' feet wide and '*b*' feet long (where '*a*' and '*b*' represent any quantity) will have an area of $a \times b$ square feet."

Even if we count *a* and *b* as being words we still have reduced the statement to 27 words and made it clearer. When we become used to algebra we can cut it even more with the aid of a diagram like this.

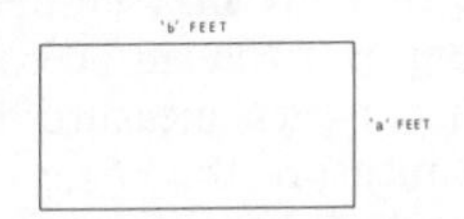

In a rectangle, width *a* feet, length *b* feet, the area = $a \times b$ square feet.

This is a very simple example and even here the use of general symbols to represent *any* width and *any* length saves a great amount of writing and explanation. Without the use of symbols in this way a calculation which can now be done in a couple of pages would begin to look like a verbatim report of a congressional session.

So if we use pebbles to represent sheep and groups of sheep we can sit in comfort in the shade and do our calculations with things which can easily be shifted from one group to another and which will stay put as long as we wish. Then when we get our answer in numbers of pebbles we can simply say that it represents the answer in numbers of sheep.

If we use symbols we can produce a general statement which is neat and concise and which can easily be applied to any particular case. Take the general statement above that a rectangle with a width of "a" feet and a length of "b" feet has an area of $a \times b$ square feet. If we have a particular case where the width is 10 feet and the length 15 feet we say simply that a equals 10 and b equals 15, therefore the area equals $a \times b$ square feet, equals 10×15 square feet, which is of course 150 square feet.

There is another way of using one thing to represent something else which is extremely useful in our modern world, where we have a lot to do with electric currents, speeds, pressures, weights, times, temperatures, and so on.

What we do in this case is to make units of length (which we can draw as lines on a piece of paper) represent units which we can't draw, such as electric current, speed, pressure, and so on.

In fact, this is what we have to do if we want to record these ideas (we can't call them "things") at all. Whatever time and speed and pressure may be they certainly are not little pointers moving over the surface of a dial.

These gadgets do not even measure time and speed and pressure, etc.; they measure the *effect* of these ideas on a suitable mechanism which in turn causes some indicator to move along or around a suitably graduated scale.

If we fail to realize this important fact we can finish by being slaves to these gadgets, and obey them instead of relying on our own common sense. The ordinary barometer, for instance, has no magical powers of seeing into the future. It has been noticed by weather observers that if the atmospheric pressure drops, the weather is likely to become stormy and if the atmospheric pressure rises the weather is likely to become fine. So the barometer is only a gadget in which changing atmospheric pressure causes a pointer to be moved around a dial. If it is taken up a mountain where the atmospheric pressure is always lower it will predict gales and rain even though it is a perfect midsummer day without a cloud in the sky. It just doesn't know any better.

Another gadget which is often treated with reverence is the lie detector. Actually this gadget has no more knowledge of what is true or false than the next door cat. What it does is measure the electrical resistance through the skin from one hand to the other of the person being questioned. It has been found that if a person is telling what he thinks is a lie any sense of guilt or embarrassment will cause his hands to sweat very slightly. This will reduce the electrical resistance. But whether he sweats from this cause, or because he gets a sudden twinge of indigestion, it is all one to the lie (or should we call it indigestion ?) detector.

If you have enough brazenness to be able to tell lies without even an unconscious sense of guilt or embarrassment the lie detector will pass you as a paragon of truth and virtue.

There is a story about a patient in a mental home who believed he was Napoleon Bonaparte. This patient was normal in many ways and realized that as long as people knew he believed he was Napoleon he would not be released. So he began to tell everyone that he had got over his delusion. On one occasion when he was being questioned with the aid of a lie detector he was asked if he was Napoleon, and he vigorously denied it. The lie detector immediately recorded that he was telling a whopping lie. According to the lie detector, therefore, he must really have been Napoleon Bonaparte after all.

So if we draw a line with units of length representing units of pressure, or speed, or temperature, or anything else we must remember that the differences in length only *represent* the differences in temperature in the same way that the pebbles represent numbers. With this in mind we will look at some of these scales and see how they can be used.

We are all familiar with thermometers in some form or

another. We use them for taking our temperature if we are sick; most kitchen ovens nowadays have a thermometer in the door; we often have one hanging in the hall beside a barometer. Most of the thermometers we use are graduated in a scale known as the Fahrenheit temperature scale. In this scale the freezing point of water is 32°F and the boiling point is 212°F. The other thermometer scale widely used in scientific work, and also the standard scale in countries using the metric system, is the centigrade, which as its name implies has a hundred divisions between the freezing and boiling points of water. Freezing point is 0°C and boiling point is 100°C. Compared with the centigrade scale the Fahrenheit scale seems very clumsy and at first sight we might wonder why sensible people ever dreamed of using such a peculiar scale. The fact is that the Fahrenheit scale was invented long before a metric system was ever dreamed of and in those days it was believed that 0°F was the coldest that anything could be. We know now of course that temperatures can drop hundreds of degrees below this, but until we decide to change to the metric system we are stuck with the Fahrenheit scale. The reason why there are 180 divisions (that is, 212-32) between freezing and boiling point is probably due to the fact that 360 divisions—the universal favorite of olden times—were too many for convenience and so 180 divisions were chosen as the next best thing.

There is no doubt, incidentally, that 360 was a favored number. There are 360 degrees in a circle; there used to be 360 days in a year. One reason for its popularity among ancient peoples was that it was so easy to divide. Out of the numbers from 2 to 20 you will find that 2, 3, 4, 5, 6, 8, 9, 10, 12, 15, 18. and 20 will all divide evenly into 360, while the only numbers that will divide evenly into 100 are 2, 4, 5, 10, and 20.

What we are concerned with, however, is the fact that, since both the centigrade and Fahrenheit scales are in common use, we often need to translate degrees centigrade into degrees Fahrenheit and vice versa. The most obvious way is to have a comparison table. Part of one is shown below.

Degrees centigrade	Degrees Fahrenheit
0	32
5	41
10	50
15	59
20	68
25	77
30	86
35	95
40	104

If we look carefully at the table we will notice several things. First, there is a constant relationship or ratio between the scales. A 5°C change is equivalent to a change of 9°F. In addition, 32°F is equivalent to 0°C.

It works out that we can convert degrees centigrade into degrees Fahrenheit by multiplying them by $\frac{9}{5}$ and adding 32. For instance, 20°C is $(20 \times \frac{9}{5})$ plus 32°F which is 36 plus 32 which is 68°F.

To convert degrees Fahrenheit into degrees centigrade we have to do exactly the opposite, that is subtract 32 and then multiply by $\frac{5}{9}$.

Another thing we can see about this particular table is that it only gives equivalents at intervals of 5 degrees on the centigrade scale and 9 degrees on the Fahrenheit scale and only goes from freezing point to boiling point. Even so, the table has quite a lot of figures. If we gave equivalents at intervals of 1 degree it would be five times larger still. We would also find that the centigrade degrees which are not evenly divisible by 5 would give odd fractions of degrees on the Fahrenheit scale. So if we also wanted to have equivalents of the Fahrenheit degrees we would need some further 150 entries in our table. The table on the next page gives some idea how bulky it would be. The bit shown covers only a twentieth part of the table between freezing and boiling point, and it still only gives us equivalents at 1 degree intervals. If we wanted equivalents at smaller intervals the table would fill pages.

Degrees C.	Degrees F.	Degrees C.	Degrees F.
0	32	$2\frac{7}{9}$	37
$\frac{5}{9}$	33	3	$37\frac{2}{5}$
1	$33\frac{4}{5}$	$3\frac{1}{3}$	38
$1\frac{1}{9}$	34	$3\frac{8}{9}$	39
$1\frac{2}{3}$	35	4	$39\frac{1}{5}$
2	$35\frac{3}{5}$	$4\frac{4}{9}$	40
$2\frac{2}{9}$	36	5	41

In an effort to get something which would be easier to use, the ingenious idea of making a "picture" of the ratio was developed. If we look at an ordinary household thermometer like the one shown below we will realize that we *see* temperature, we don't hear it or feel it or taste it—even though we may feel the effect of it! In this sense the

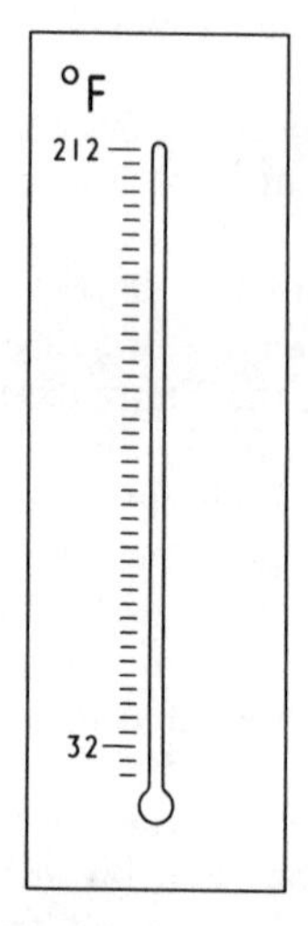

(a) ORDINARY THERMOMETER

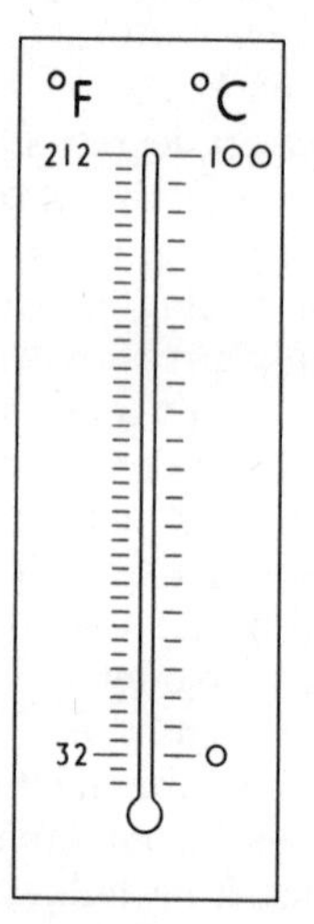

(b) THERMOMETER WITH TWIN SCALES

thermometer already gives us a picture of the actual temperature. When the centigrade scale began to be widely used, some thermometers were made with the Fahrenheit degrees shown on the other side. Thus it was possible to read both scales on the one thermometer.

This means, since both scales refer to the same thermometer, that at every level the markings on each scale represent the same temperature.

If for instance the temperature is at freezing point the mercury will be at a certain level. At this level the Fahrenheit scale must read 32°F, and the centigrade scale must read 0°C., which is the equivalent reading in the centigrade scale. So if we take the thermometer away we are left with the two scales, and each mark on one scale is exactly opposite the position on the other scale which represents the same temperature as is shown below.

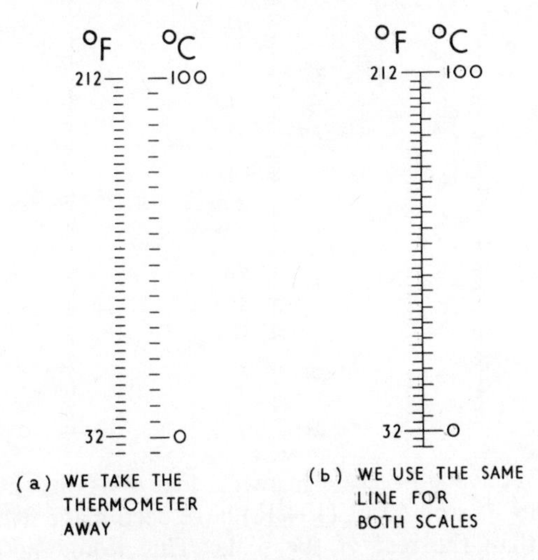

(a) WE TAKE THE THERMOMETER AWAY

(b) WE USE THE SAME LINE FOR BOTH SCALES

We can go one step further and put the two scales on different sides of the same line. Then we have what is called a conversion scale and it is probably the best and neatest way of expressing this kind of relationship.

The practical conversion scale shown below would give all the necessary information for converting from degrees F. to degrees C. and vice versa between freezing and boiling points. It also illustrates two extra advantages of this type of conversion scale. To make it more compact, it can be cut into convenient lengths and the pieces placed side by side without affecting its usefulness, and also we can "stretch" any part of the scale if we wish to be able to read any particular part of the range to a greater degree of accuracy.

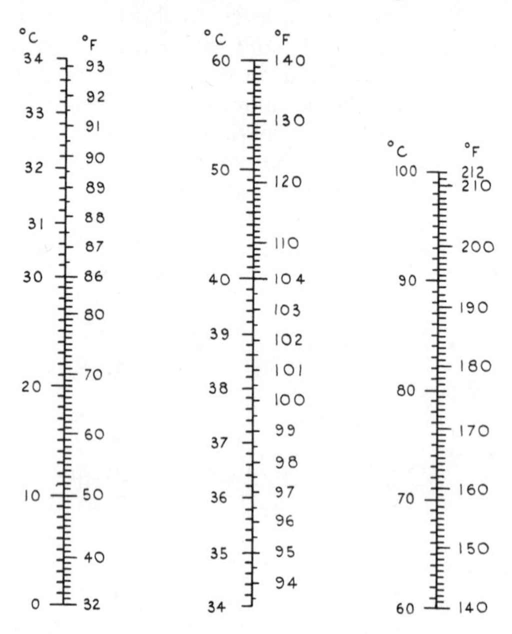

In the conversion scale shown, the divisions between 30°C. (86°F.) and 40°C. (104°F.) have been made five times longer than the rest of the scale. This would make the conversion scale suitable for medical use where a high degree of accuracy at body temperatures is required.

The conversion scale has been mentioned at some length because it is used, in some form or another, in almost every case where quantities are shown in pictorial form. Take, for instance, a scale such as we see at the bottom of a map. It usually looks something like this:

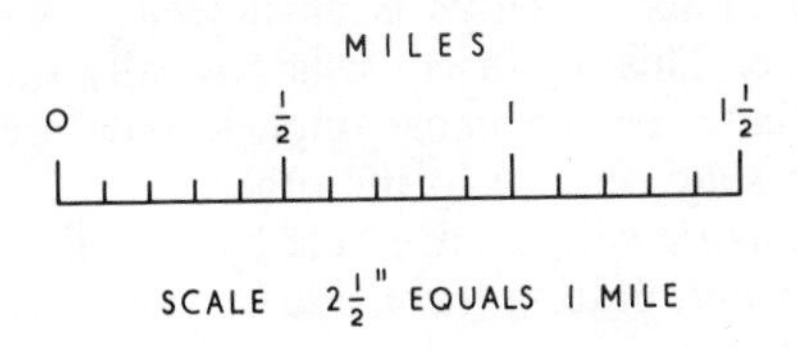

Actually it is a conversion scale with the other half not drawn. The full scale would look like this:

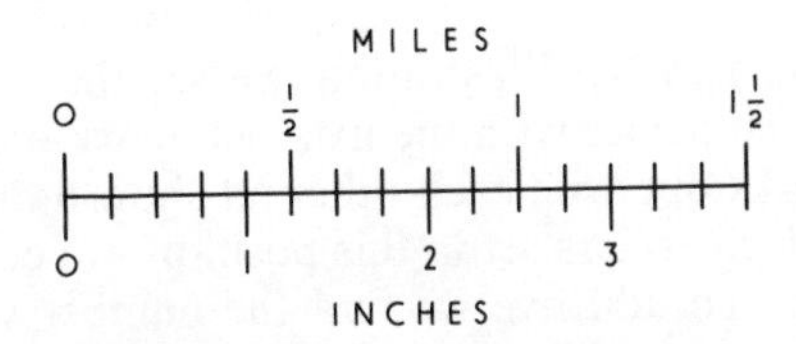

We don't need the bottom half because we are told that the distance marked 1 mile is $2\frac{1}{2}''$, and so, instead of measuring distances on the map in inches and then looking up the conversion scale, it is just as easy to use the top scale and get the real distance in miles straight away.

The conversion scale is the first and simplest example of "abacus automation"—gadgets which do our calculating for us without our having to do any calculations at all or even put pen to paper. Conversion from one set of units to another and map scales are the most common types, but we can use conversion scales to show numbers and their square roots or numbers and their squares or numbers and their logarithms. etc. We could also show the distance travelled by a car and the number of gallons of gas used (and the cost) or the time for which an electrical appliance was used and the cost.

Many weighing machines in grocers' shops have sets of conversion scales built in. The top scale reads the weight and the other scales tell the total cost at 50¢ per pound or \$1.00 per pound or \$1.50 per pound and so on.

Another automated gadget we can use if we want to avoid doing calculations ourselves is an ingenious gadget called the slide rule. This again consists basically of two scales but in this case, as the name implies, they are not fixed, but one can slide relative to the other.

We can make a simple slide rule by taking two ordinary school rulers and putting them side by side as shown in the sketch below.

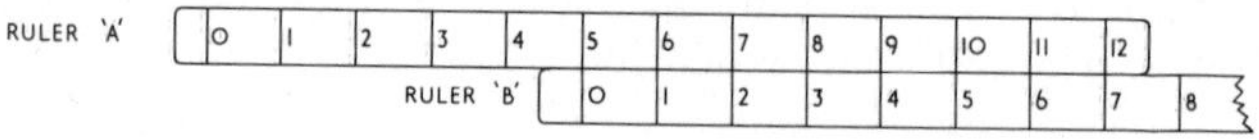

RULER 'A' 0 1 2 3 4 5 6 7 8 9 10 11 12
RULER 'B' 0 1 2 3 4 5 6 7 8

If, as shown in the illustration, we set the ruler "B" so that its zero coincides with the five-inch mark on ruler "A" then the marks opposite each other all show a difference of five. So, with the rulers set in this position, we could add or subtract five. To add five we find the number we want to add to five on ruler "B" and then get the answer at the mark opposite on ruler "A". To subtract we find the number we want to subtract five from on ruler "A" and then the answer is given by the mark opposite it on ruler "B". To add or subtract by any other number we merely slide the rulers until the zero on one rule is opposite the figure we want to add or subtract on the other.

This adding and subtracting rule isn't much use unless it is very long or has a great number of fine divisions, because we can do most simple additions and subtractions mentally. But it is the basis for an extremely useful and interesting gadget, a slide rule which multiplies and divides.

In the previous section on logarithms we saw how numbers could be multiplied together by adding their logarithms.

Now if we look at the scales on our adding slide rule we will find they are linear. This means that the distance along from 0 to 1 is the same as the distance from 1 to 2 and from 2 to 3 and so on. From 0 to 6 is twice as far as from 0 to 3. Another way of putting this would be to say that every unit of quantity is represented by one unit of distance along the line.

If we make a scale where these distances are proportional to the logarithms of the numbers instead of to the numbers themselves we will have a scale which can be used for multiplication instead of addition. If you can't quite understand this, forget it for a moment and look at the sketch below.

-1 0 1 2 3

This is a perfectly straightforward linear scale with numbers. Now let us suppose that these numbers are logarithms. We will remember from the previous section that

3	is the	logarithm	of	1000	because	10^3	equals	1000
2	,,	,,	,,	100	,,	10^2	,,	100
1	,,	,,	,,	10	,,	10^1	,,	10
0	,,	,,	,,	1	,,	10^0	,,	1
−1	,,	,,	,,	0.1	,,	10^{-1}	,,	$\frac{1}{10}$ (or 0.1)

With this information we can draw this line again and make it into our old friend the conversion scale, calling the numbers on top logarithms and putting underneath the actual numbers they represent, like this:

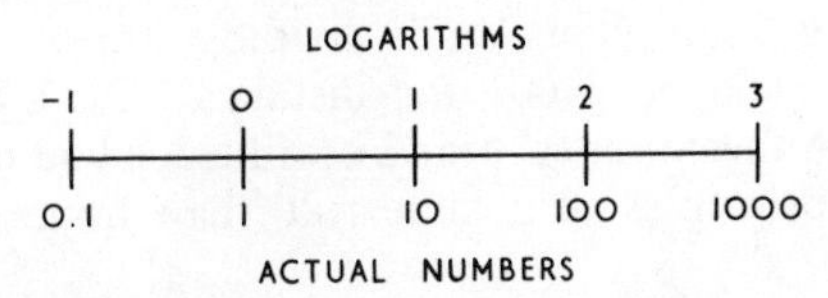

We have now got the scale we mentioned above where the distances are proportional to the logarithms of the numbers and not to the numbers themselves.

You can now see the difference. In an ordinary linear scale it would be twice as far from 0 to 20 as it would be from 0 to 10. But in the logarithmic scale it is twice as far from 1 to 100 as it is from 1 to 10. This is where the multiplication comes in. If we make two identical scales (leaving off the top part showing the logarithms because we don't need it) and slide one along until the 1 coincides with the 10 mark on the other scale we would have something like this:

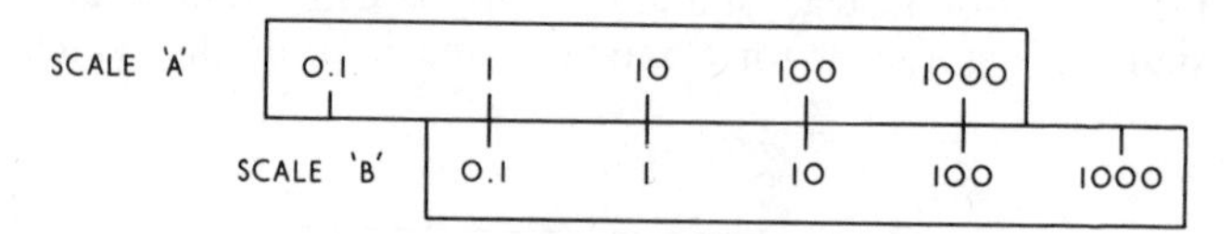

And so we are back to our adding rules but this time they multiply and divide. All the quantities shown on scale "A" are equal to the corresponding marks on scale "B", multiplied by 10, and all those on scale "B" equal to the corresponding marks on scale "A" divided by 10.

At this stage you should make a couple of scales out of cardboard and mark them at inch intervals with a ruler. Then label a division at about the center as 1, the next ones to the right as 10, 100, 1000, and so on and the ones to the left as 0.1, 0.01, 0.001, and so on. Something like this:

0.00001 0.0001 0.001 0.01 0.1 1 10 100 1,000 10,000 100,000

Now try multiplying and dividing various numbers. What is 10,000 × 0.001 for instance? Or 1 divided by 0.00001? A few minutes' practice will teach you more about the basic principles of a slide rule than hours of careful reading.

If it is to be of practical use, however, the slide rule must have a scale with markings representing all the other figures as well. It is not at all difficult to do this provided we have a table of logarithms handy.

The illustration below shows what a normal logarithmic scale looks like. The scale shown goes from 1 to 10 (this is called a decade), but if we want to go from 10 to 100 or 0.1 to 1 the length and spacing will be exactly the same over again. A scale from 1 to 100 for instance would be made up of two identical decades.

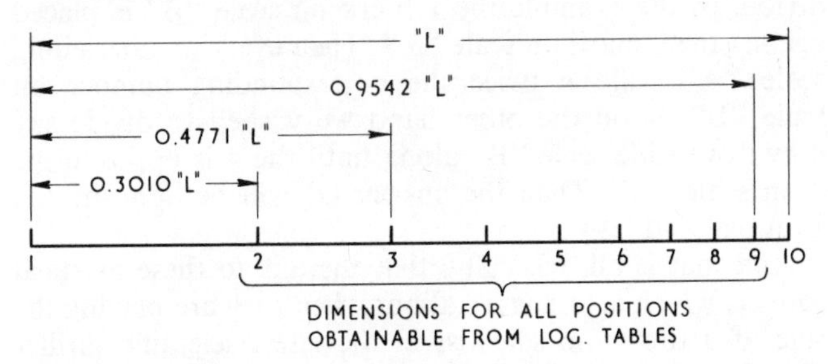

To make up our logarithmic scale we choose any convenient length (called *L* above) and mark 1 at the left-hand end and 10 at the other.

To find where to put the numbers 2, 3, 4, etc., and any other divisions we want to put in, we look up the logarithm of these numbers and multiply them by the decade length which we have called *L*.

For example, the logarithm of 2 is 0.3010 so 2 will be at a mark which is 0.3010 of the total length *L* measured from the left-hand end as shown. The logarithm of 3 is 0.4771 so 3 will be 0.4471 of *L*, and so on.

We can now see why 1 is at the zero mark of the scale, because the logarithm of 1 is zero and thus 1 will be at a mark "zero times *L*" along the scale, which of course is at the zero mark. Similarly the logarithm of 10 is 1 and so 10 must be at a mark which is one times *L* or in other words at *L* itself which is the end of the scale.

By now you may have realized that the one-decade logarithmic scale shown above is just an enlarged slice of the four- and ten-decade logarithmic scales shown on page 86 and can be used in exactly the same way. The sketch below shows a pair of scales side by side:

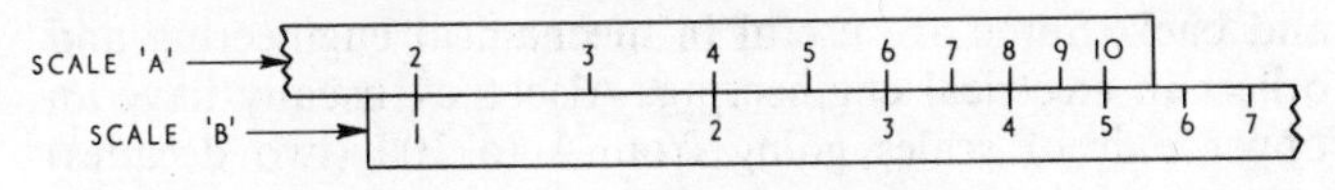

Using exactly the same basic principle as we used in the adding and subtracting slide rule we can now multiply and divide. In the example the 1 mark on scale "B" is placed against the 2 mark on scale "A". Then every number along scale "A" will be twice the corresponding number on scale "B". If on the other hand we wished to divide say 8 by 4 we slide scale "B" along until the 4 is opposite the 8 on scale "A". Then the answer (2) will be opposite the 1 on scale "B".

And that is all, basically, that there is to these mystical gadgets which actors wave about when they are playing the role of a top atom scientist in the latest scientific thriller film. But apart from their use by film stars and engineers and scientists (engineers and scientists do actually use them) a slide rule can be quite handy about the house. If you want to make a quick check of Junior's arithmetic homework without going to the trouble of working it all out yourself, if you want to find how many yards of concrete are needed for the front path, the number of gallons of paint needed to do the house, or yards of paper to do the walls and ceilings, or the cost of the new fence, you can do it on a slide rule without any mental effort. It is the greatest help where there are a lot of fairly simple multiplications and divisions involved and the answer need not be accurate to more than about three figures. The actual accuracy of the slide rule depends on the number of divisions one can squeeze in and this in turn depends on the length of the decade. Most slide rules are about a foot long but some monsters five or six feet long have been made for special scientific jobs. It is not entirely true, however, that the social standing of a scientist is proportional to the length of the slide rule he uses, though there may be a tendency that way.

The commercial slide rules—in addition to the logarithmic scales—usually have various conversion scales on the sides and back. Some are useful in mechanical engineering and others in electrical engineering. Also they usually have an upper pair of scales going from 1 to 100 (two decades) and a lower pair going from 1 to 10 (1 decade). The advantage of this is that the figures on the top pair of scales are the squares of the figures on the bottom.

The sketch below shows the four basic scales on a slide rule. The two center scales are on a sliding section as shown.

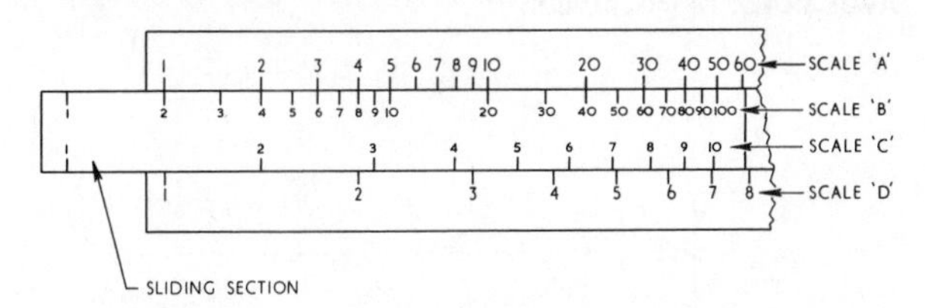

The sketch below shows the rule with the center sliding section removed and we can see clearly how the numbers on scale "A" are the squares of the numbers directly below them on scale "D"

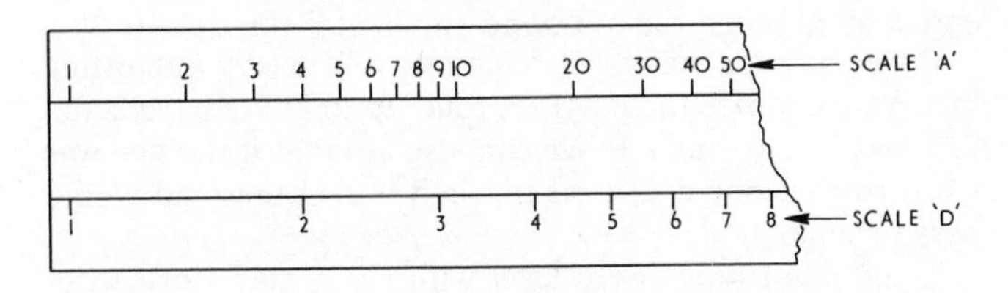

The slide rule is a fascinating gadget as will be discovered if you buy, borrow, or better still make one. One can play with it for hours and it is at least as useful as doing crossword puzzles.

In addition to conversion scales and slide rules there is another interesting and useful device which can save a lot of dreary calculating. This is the nomograph, and it uses some of the principles of the conversion scale and some of those we find in the slide rule.

If you have ever watched two children playing on a see-saw you will have some practical knowledge of the geometric principles on which a nomograph relies. The see-saw is a straight plank pivoted at a fixed point in the center. When either end goes up, the other end has to go down by an equal amount. We could measure this by having a ruler standing upright at each end of the plank.

The sketch below represents a mathematical version of this see-saw. The scales at either side are equal, and the pivot point is in the center.

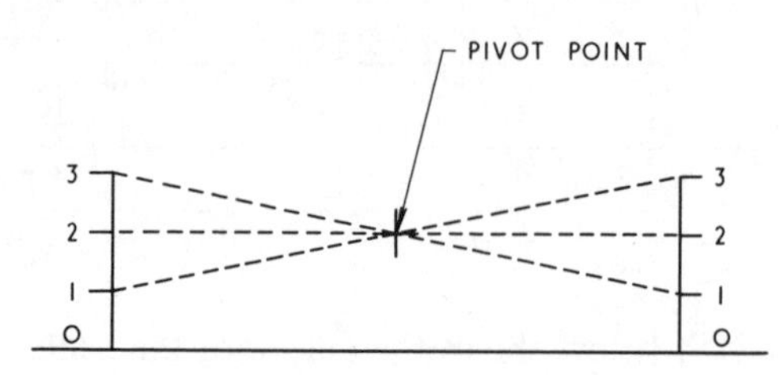

The three dotted straight lines represent the see-saw in three different positions. When it is level both ends of the line are at 2. When the left-hand end is at 1 the right-hand end is at 3. When the left-hand end is at 3 the right is at 1.

If we consider this for a minute we find a very interesting fact. In every case these figures total up to 4. We have 2 and 2, 3 and 1, and 1 and 3. We can also see that if the line was tilted so that one end went through 0 the other end would pass through 4.

If the pivot point were level with the 1 mark instead of the 2 marks we would have a similar effect, as shown in the sketch below.

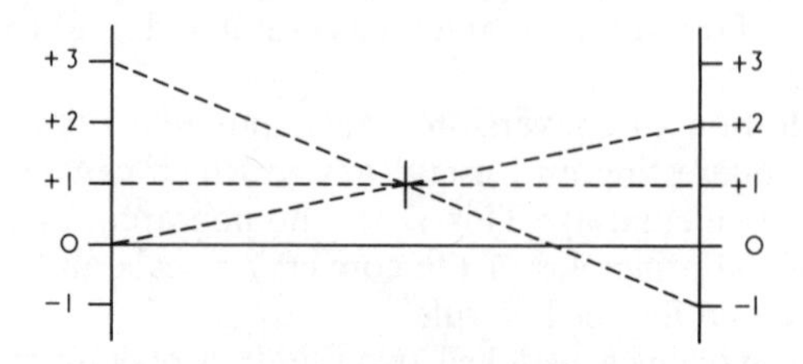

This time we see, even if we extend the side scales past zero to -1, that the figures on the two scales always add up to 2. We have 1 plus 1, 2 plus 0, and 3 plus (-1).

If the pivot point were level with the zero marks we would find that the sum of the markings on the two side scales always added up to zero. Now suppose we put a third vertical scale between the two side ones. The point level with the 1 marks on the side scales we can label 2, the point level with the 2 marks on the side scales we can label 4 and so on. We will then have our three vertical scales marked as shown below.

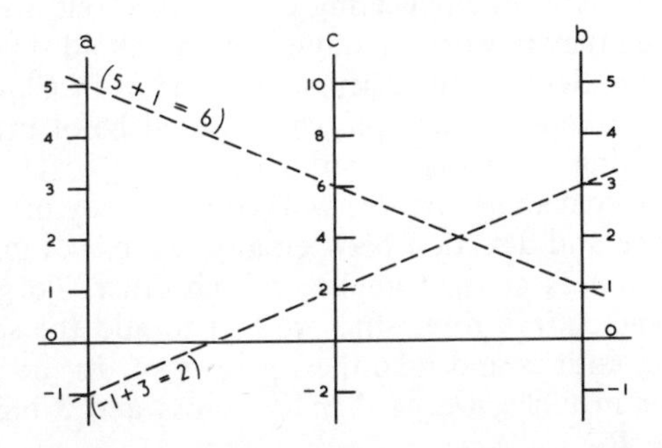

If we let the left-hand scale markings represent one quantity (we will call it *a*) and the right-hand scale markings represent a second quantity (we will call this *b*) then the center scale will represent a quantity (we will call it *c*) which is the sum of quantities *a* and *b*. So the nomograph will give us the value of *c* when we choose values for *a* and *b*. If for instance *a* is 5 and *b* is 1, we can lay a ruler on the nomograph so that its edge passes through mark 5 on *a* and mark 1 on *b* and we will find that it also passes through mark 6 on *c*, which is our answer.

The other dotted line on the sketch shows the position of the ruler for getting the answer to —1 plus 3. Here we see that the ruler cuts the center scale at 2, which is the answer.

For subtraction we simply use the center scale and one of the side ones, and read the answer on the remaining side scale, because if *a* plus *b* equals *c* then *a* must equal $c-b$ and *b* is equal to $c-a$.

In this simple form, of course, the nomograph is not

much use because we could do these kinds of addition mentally. However, when we use conversion scales, our nomograph can become very valuable.

There is an almost endless variety of types of conversion scales. One useful one is where a unit of measurement is converted into a unit of money. Another is our old friend the logarithmic scale, by which we can convert the nomograph from one which adds to one which multiplies. Nomographs are useful in engineering design work but are more bother than they're worth in doing the household shopping.

But in the light of our discussion on maps in Chapter 1 we could make a nomograph which would be of practical use in working out distances on maps.

You will remember that we worked out a way of finding the distance and direction between any two places in terms of two distances at right angles to each other. To get the actual distance in a direct line we had to add the squares of these quantities and take the square root. In the sketch below, for instance, we go 4 miles across and 3 miles up from *A* to *B*.

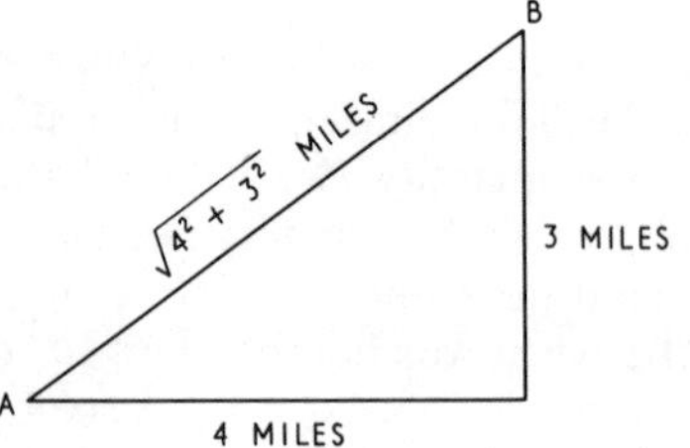

The direct distance is then the square root of 4^2 plus 3^2.

So A to $B = \sqrt{4^2+3^2} = \sqrt{16+9} = \sqrt{25} = 5$ miles.

The distance from the Cock and Bull to the Opera House on the map in Chapter 1 was $2.4 \times i\,1.8$ miles, which looked like this:

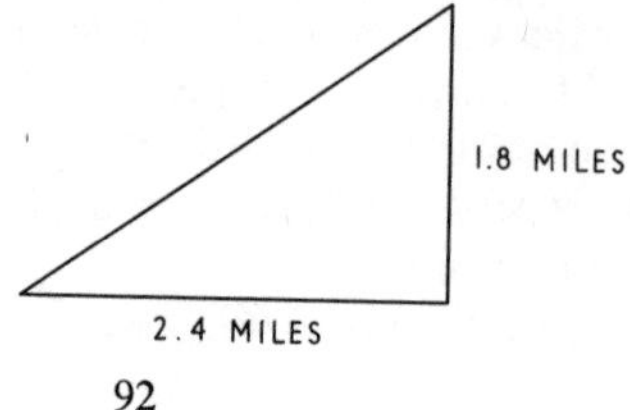

The direct distance here was:

$\sqrt{2.4^2+1.8^2} = \sqrt{5.76+3.24} = \sqrt{9.00} = 3$ miles

These calculations can be entirely eliminated by using a nomograph. Let us see how a practical one can be constructed.

We know that what we are trying to do is to find the length of the longest side (or hypotenuse) of a right-angled triangle. We know it is the square root of the sum of the squares of the other two sides. Let us try to put this in a clearer and shorter way. We will begin by drawing a right-angled triangle and saying that its two short sides are a units long and b units long, where a and b are any values we like.

The triangle looks like this:

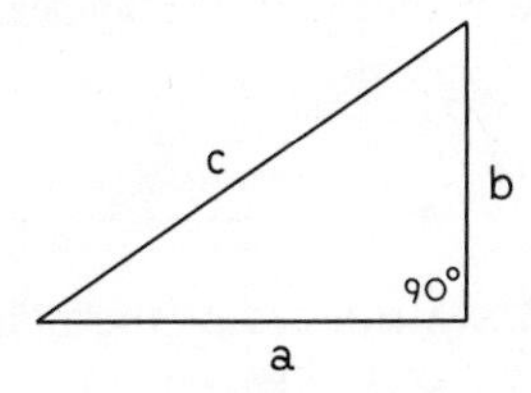

From what we have said above we know that the length of the third side "c" must be equal to $\sqrt{a^2+b^2}$. Or, putting it more simply: $c = \sqrt{a^2+b^2}$.

Having got this fairly simple general formula it is possible to make it even simpler still. We will remember that if the square root of any quantity is squared it becomes the quantity itself. For instance $\sqrt{9}$ is 3; and $\sqrt{9^2}$ is $\sqrt{9}\times\sqrt{9}$, or 3×3 which is equal to 9.

In the same way, if we square $\sqrt{a^2+b^2}$ we get

$$\sqrt{a^2+b^2}\times\sqrt{a^2+b^2}$$

which is a^2+b^2.

Since $c = \sqrt{a^2+b^2}$ we can square both of the equal sides of the equation and the result will still be equal. So we can say that if $c = \sqrt{a^2+b^2}$

then $c^2 = a^2+b^2$

If we wanted to we could work out this last equation on the simple addition type of nomograph illustrated on page 91. Take a case where a is equal to 4 and b is equal to 3. Then $a^2 = 4 \times 4 = 16$, and $b^2 = 3 \times 3 = 9$. So $a^2 + b^2 = 16 + 9 = 25 = c^2$.

The trouble with this is that we have to do the squaring and take the square root—the harder parts of the job—ourselves. Once we know that a^2 equals 16 and b^2 is 9 we can add 16 plus 9 without the help of the nomograph.

Fortunately, however, there is a fairly simple way out of the difficulty.

What we have to do is to enlist the help of the conversion scale. In the first nomograph we had a scale where the spaces represented the plain numbers. This time we want one where the spaces represent the *squares* of the numbers. Like this:

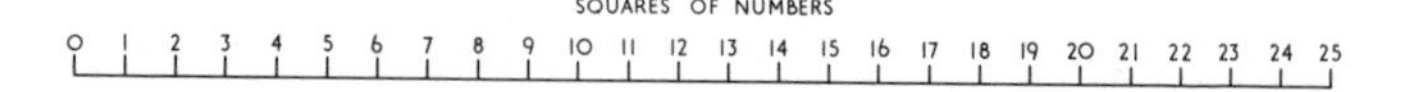

Then opposite these squares of the numbers we put the numbers themselves. Like this:

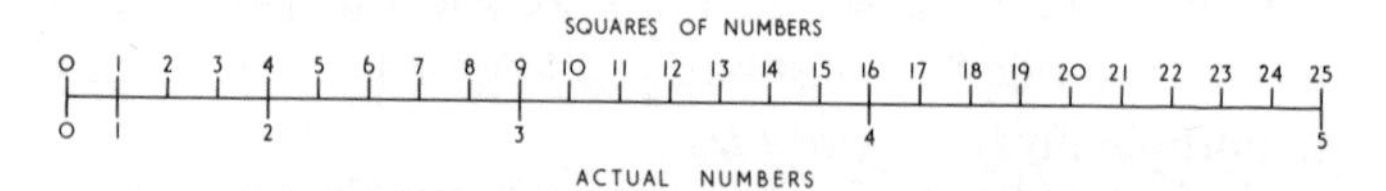

You can see that the distances between the numbers are no longer even but are proportional to the *squares* of the numbers. So now when we add 3 plus 4 on the scale we are actually adding 3^2 plus 4^2

We could, by the way, use two of these scales in the form of a slide rule. If you are interested you could mark them out from the table on page 96. The two scales would be the same as that shown above and the calculation giving 3^2 plus 4^2 equals 5^2 is shown below.

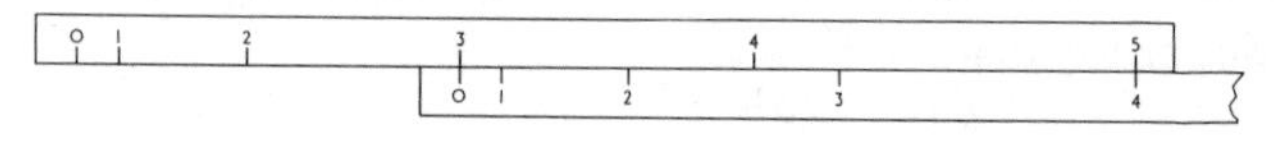

Zero on the bottom scale is set against 3 (representing 3^2) on the top scale. Then we find that 4 (representing 4^2) on the bottom scale is opposite 5 (representing 5^2) on the top scale. So we can see that our simple adding slide rule can be used for multiplying, adding squares, and all sorts of other jobs by using suitable conversion scales.

The same thing applies to the nomograph, only more so. Where the two scales on the slide rule have to be the same, the two outer scales on the nomograph can each be different. We could, if we wanted, make a nomograph of a plus b^2 equals c^2, or $3a^3$ plus $\frac{1}{2}b$ equals c^4, or any other combination we liked, just by using appropriate conversion scales. However, since the example we have chosen is a useful one, we will stick to the a^2 plus b^2 equals c^2, converted back to the form, $c = \sqrt{a^2+b^2}$.

The first job is to make a table of figures from which we can draw our conversion scales. This is simply a list of the values we want to mark on our scales with the values of their squares opposite them. Like this:

Number	Square of the Number
0	0
1	1
2	4
3	9

What makes it tedious is that to get a useful working nomograph we have to get all the in-between figures and their squares. For this nomograph it will be good enough if we get a mark for every $\frac{1}{10}$ such as 1.1, 1.2, etc. Finally we will remember that in our addition nomograph the center scale spacing was half that of the outside scales. This still applies so we will add a third column to our table giving the distances for the center scale, shown on page 96.

To construct our nomograph we draw a horizontal base line and a center vertical line and two outer vertical lines at equal distances from the center line. The actual distances can be anything that is convenient.

Number	Square for Scales *a* & *b*	½ Square for Scale *c*	Number	Square for Scales *a* & *b*	½ Square for Scale *c*
0	0	0	4.0	16.00	8.00
0.5	0.25	0.125	4.1	16.81	8.405
1	1	0.5	4.2	17.64	8.82
1.1	1.21	0.605	4.3	18.49	9.245
1.2	1.44	0.72	4.4	19.36	9.68
1.3	1.69	0.845	4.5	20.25	10.125
1.4	1.96	0.98	4.6	21.16	10.58
1.5	2.25	1.125	4.7	22.09	11.045
1.6	2.56	1.28	4.8	23.04	11.52
1.7	2.89	1.445	4.9	24.01	12.005
1.8	3.24	1.62	5.0	25.00	12.50
1.9	3.61	1.805	5.1	26.01	13.005
2.0	4.00	2.00	5.2	27.04	13.52
2.1	4.41	2.205	5.3	28.09	14.045
2.2	4.84	2.42	5.4	29.16	14.58
2.3	5.29	2.645	5.5	30.25	15.125
2.4	5.76	2.88	5.6	31.36	15.68
2.5	6.25	3.125	5.7	32.49	16.245
2.6	6.76	3.38	5.8	33.64	16.82
2.7	7.29	3.645	5.9	34.81	17.405
2.8	7.84	3.92	6.0	36.00	18.00
2.9	8.41	4.205	6.1	37.21	18.605
3.0	9.00	4.50	6.2	38.44	19.22
3.1	9.61	4.805	6.3	39.69	19.845
3.2	10.24	5.12	6.4	40.96	20.48
3.3	10.89	5.445	6.5	42.25	21.125
3.4	11.56	5.78	6.6	43.56	21.78
3.5	12.25	6.125	6.7	44.89	22.445
3.6	12.96	6.48	6.8	46.24	23.12
3.7	13.69	6.845	6.9	47.61	23.805
3.8	14.44	7.22	7.0	49.00	24.50
3.9	15.21	7.605			

THESE ARE NOT REQUIRED

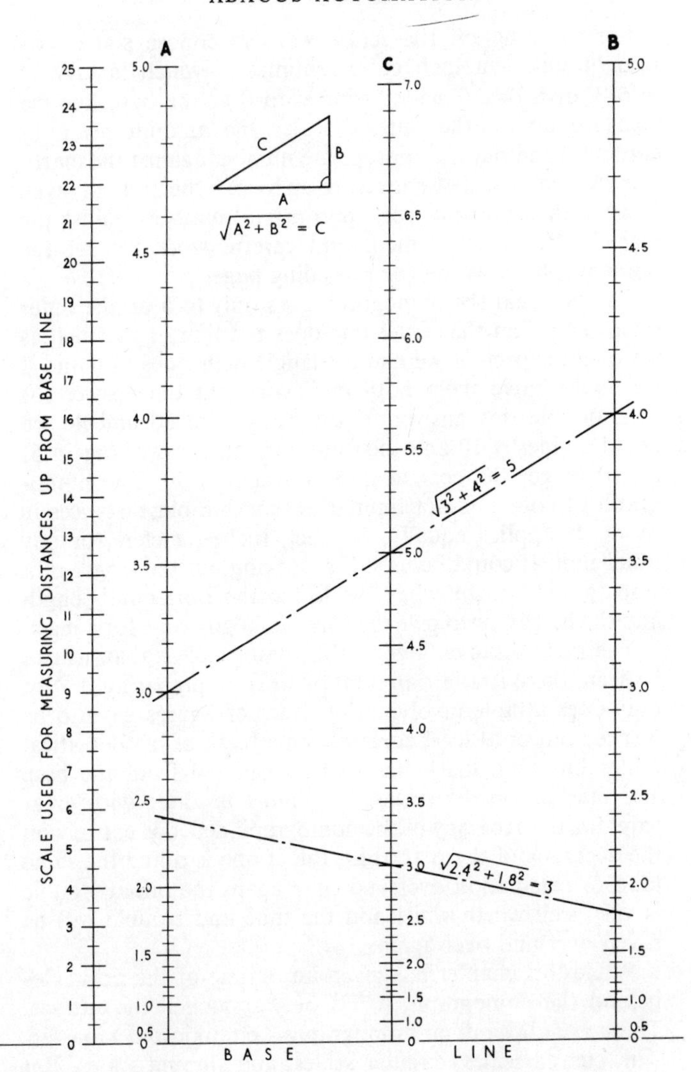

SCALE USED FOR MEASURING DISTANCES UP FROM BASE LINE
A
B
C
$\sqrt{A^2 + B^2} = C$
$\sqrt{3^2 + 4^2} = 5$
$\sqrt{2.4^2 + 1.8^2} = 3$
BASE LINE

For marking off the scales we also choose some convenient unit—an inch or a centimeter—which is divided into $\frac{1}{10}$ divisions. Then starting with 0 at the base line we measure up on the outside scales the amount given in column 2 and put the appropriate number against the mark. For the center scale we measure up from 0 the amount given in column 3 and put the appropriate number against the mark. After much patient and careful work we get the nomograph shown on the preceding page.

The fact that the nomograph goes only to 5 on the outer scales and 7 on the center one does not limit it as much as we might expect. If we had a triangle with sides of 6 and 8 we could halve them both and work out the answer (5) then double this answer. If the sides were 30 and 40 we could divide by 10 and then multiply our answer (again 5) by 10 to get the correct answer which is 50. The nomograph of course is not limited to working out distances in miles. It applies equally, for feet, inches, meters, or any other unit. It could be used for working out the length of a sloping roof beam when we know the horizontal length and the height, or to give the length of a guy rope for a mast.

The main disadvantage of the nomograph is that, as can be seen, there is a fair amount of work in preparing it. Not only does a table involving hundreds of figures have to be worked out or at least copied from a book of mathematical tables but each mark has to be measured from the base and marked on the scales. This must be done with great care, for the accuracy of the nomograph depends entirely on the accuracy of the markings. But if one is doing the same kind of calculation over and over again the initial trouble is very well worth while and the time and trouble will be saved over and over again.

While this chapter has given an outline of the principles behind the nomograph, it has only scratched the surface. There are all kinds of nomographs—complicated monsters with curved scales, circular scales, and sloping scales. But they all work on much the same principle and it is hoped that we have managed to remove these gadgets from the category of inexplicable and magical devices and shown that they are tools of trade as understandable, if not perhaps as generally useful, as a hammer, a spade, or a wheelbarrow.

EXERCISE

1. Make yourself a simple slide rule along the lines explained in the section. You'll be surprised how much fun you can have with it and how much you'll learn. Two strips of cardboard are all that you will need.

5. GRAPHS...

The pictures for the page atone.
Pope.

A graph can best be described as a mathematical picture. It is a means of conveying information quickly and easily through shape, and for this reason is being used more and more in modern civilization. In this chapter we will investigate the stuff that graphs are made of and of what use or otherwise they can be.

At the beginning of this book a distinction was drawn between the sheep and cow mathematics of things and the mathematics of ideas—length, direction, weight, and other kinds of measurement. In this chapter you will find yet another kind of mathematics, the mathematics of change and comparison. This is mainly what graphs are concerned with. Again there is no hard and fast dividing line and the old symbols and expressions are still largely used; but many of them have a slightly different meaning. As we have already encountered this kind of thing there is no reason why it should worry us unduly.

One of the foundation stones in this mathematics of change and comparison is the conception known as a ratio. We could roughly describe it as a division with a difference.

We have already had a look at division as it applies to sheep and cow mathematics and also to the mathematics of measurement. We saw that division had a very limited application in sheep and cow mathematics and a much wider one in the mathematics of measurement, where negative and fractional answers have a practical meaning. In the simplest kind of division we divide a group of things into several smaller groups and that is about all, but in the mathematics of measurement we can divide ideas such as the idea of length into as many parts as we like no matter how small they become. We can also divide, in some cases, one kind of unit by another. For instance, we can not only divide 200 square feet by 20 to get 10 square feet but we can also divide 200 square feet by 20 *feet* and get 10 *feet* as our answer. What we are saying in this second case is that if we have a rectangular shaped area of 200 square feet—a room, for instance—and it is 20 feet long, then it must be 200/20 feet, or 10 feet, wide.

But though we can divide square feet by feet we couldn't divide an amount of weight or a tension by a length and get an answer which means anything in this kind of mathematics. In the mathematics of change and comparison, however, we can do it and get very useful and practical answers. Suppose we had a piece of rope and discovered that it stretched two feet when a pull of 200 pounds was applied to it. In this new mathematics we can divide the 200 pounds by the 2 feet and say that this particular piece of rope has an average pull of 100 pounds per foot stretch. Or we can divide the 2 feet by the 200 pounds and say that the average stretch of the rope is 1/100 of a foot per pound pull. The stretch of a piece of rope depends in practice on many factors; the type of rope, the length, the diameter, whether it is damp or dry, etc., as well as on the pull applied. For the sake of simplicity, however, we are only considering the pull.

You will realize that both answers amount to exactly the same thing. Likewise we can, if a car travels 100 miles in four hours, divide the 100 miles by four hours and say the car travels at an average speed of 25 miles per hour, or we can divide the 4 hours by the 100 miles and say the car takes an average of 1/25 hour per mile it travels. And again we are saying the same thing in two different ways. In the kinds of mathematics we mentioned earlier, 100 divided by 4 is a very different thing from 4 divided by 100, whichever way we look at it.

The basic difference is that in the mathematics of change and comparison we don't finish with a single simple answer like a number or a distance or an area. We finish with the expression of a relationship between two things. And in such a case there are always two ways of expressing it. If we are comparing the weights of wood and lead for instance we can say that wood is lighter than lead and we can also say that lead is heavier than wood.

This expression of a relationship between two things is the conception known as ratio and we can now see why it was described as a division with a difference. This new and broader mathematical conception is essential for dealing with the type of problems we come across in the modern world. Speed is a pure ratio, a relationship between time and distance. Birth and death rates are ratios. Our incomes are not just a matter of how much money but a relationship between money and time. If you don't believe this ask yourself whether an income of a hundred dollars a year is the same as an income of a hundred dollars a week. Both are a hundred dollars per . . . only the time is different. Interest rates are again ratios, in this case double-barrelled ratios. The interest rate is so much *per* hundred dollars *per* year (or month if you are unlucky).

Before going into too much detail it might be helpful if we make some classification of the different kinds of ratios we are likely to meet and explain their differences.

Firstly there are what we know as conversion ratios. These are a half-way house between the mathematics of measurement and the mathematics of change and relationship. Although they express a relationship or ratio they are not a ratio between two different things, but a ratio between two different *ways* of describing the same thing. If we want to change pounds into dollars (both different ways of describing quantities of money) we look up tables or charts which tell us how many dollars we get per pound or how much of a pound we get per dollar. When we convert feet into meters, or miles into kilometers, or temperature in degrees Fahrenheit into temperature in degrees centigrade we are dealing with this kind of ratio. The method of expressing this kind of ratio by the use of conversion scales has already been dealt with in Chapter 4.

In Chapter 4 we saw what a conversion scale was and

how it was used. We saw how a typical conversion scale could be developed from a thermometer having a centigrade temperature scale on one side and a Fahrenheit temperature scale on the other; and how the thermometer could be removed and the two scales placed on either side of a single line, like this:

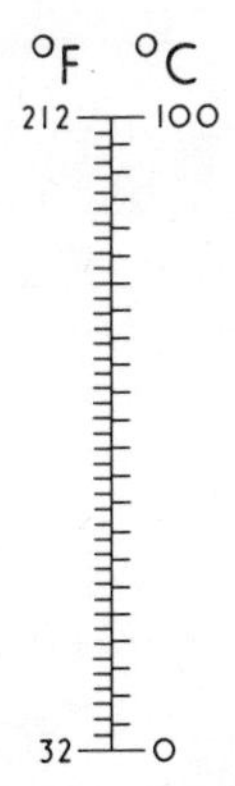

Suppose, however, that instead of placing these scales side by side we placed one at right angles to the other, like this:

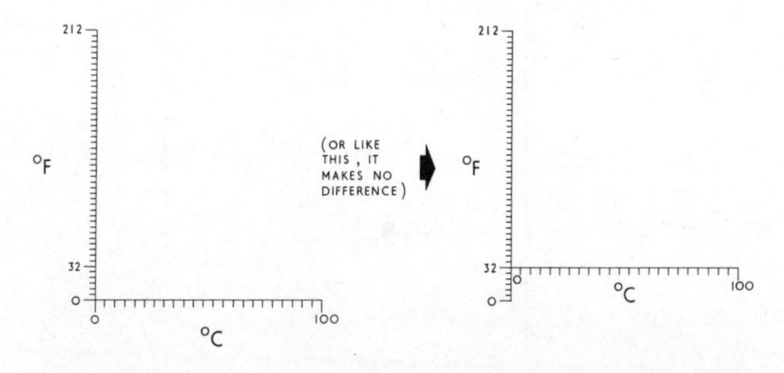

The two scales then do almost exactly the same job as the map references we discussed in Chapter 1. With their use we can find points which represent the information given in the conversion table. We know, for instance, that

100 degrees C. equals 212 degrees F., for both represent boiling point. If we go across to the right from the 212 degrees F. level and also up from the 100 degrees C. level we find a point which must be on our graph. Like this:

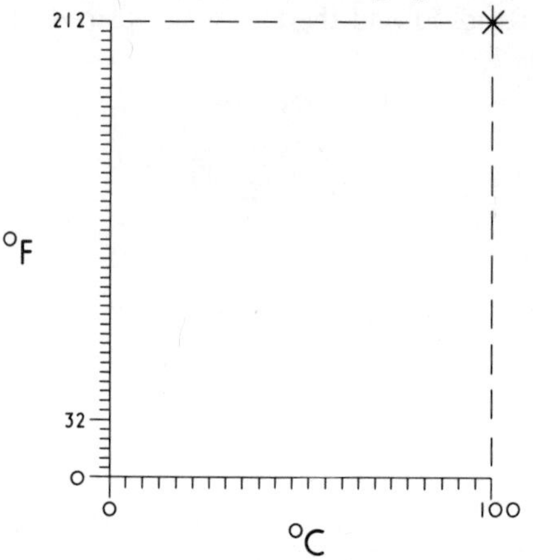

Taking another equivalent from the table we see that 40 degrees C. equals 104 degrees F. We can find this point also on the graph like this:

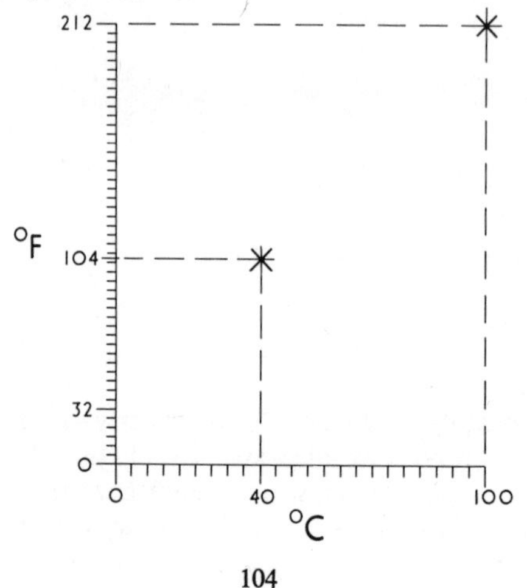

And of course 32 degrees F. equals 0 degrees C. and this is represented by the point at the edge of the vertical scale.

If we go on plotting equivalent points we will find that they all lie on a straight line. This is because the ratio between degrees F. and degrees C. is a fixed quantity over the whole scale. We can draw this straight line as shown below.

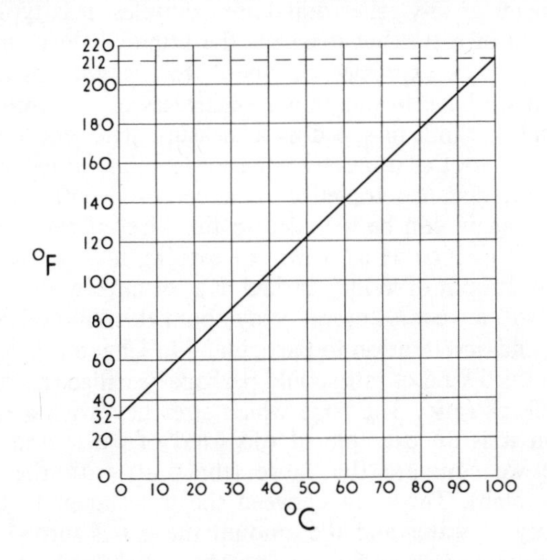

We can then pick any value of degrees F. and look along the level until we reach the line. Then we go vertically down and we will find the equivalent value in degrees C. To convert degrees C. into degrees F. we reverse the procedure, going up from the degrees C. scale and across to the degrees F. scale.

As far as conversion ratios are concerned, however, the graphical representation does not give us any more useful information than does the conversion scale illustrated in Chapter 4, except that it tells us we have a "straight line" or "linear" ratio. In addition, it is more difficult to read and occupies more space. So, in practice, for this kind of job the conversion scale is usually chosen.

Leaving conversion ratios, we come to a second kind of ratio with which we are all familiar. Many of the ratios in this category are so useful in practical everyday life that they have become units of measurement in their own right.

Speed in miles per hour (or feet per second, etc.) and pressure in pounds per square inch (or grams per square centimeter or any other units) are examples. It is typical of this kind of ratio that it means the same whatever units it happens to be expressed in. Speed, for instance, is a conception we have in our minds regardless of the units, and we tend to think of speed as a measurement—more as we think of a foot as a measurement of length—rather than as a certain distance travelled in a certain length of time. Income again can be included in this kind of ratio. Again we think of income as a *rate* of earning money, not as a certain amount of money earned in a certain length of time.

We will have a lot more to say about this kind of ratio—speed and acceleration in particular—in Chapter 7.

The third kind of ratio could perhaps be called the "cause-and-effect" ratio. The rope which stretches when a load is put on it is an example of this kind of cause and effect where we compare the cause (the pull) with the effect (the stretch). The ratio between the time taken to boil a quantity of water and the amount the gas is turned up is another cause-and-effect ratio. The relationship between smoking and lung cancer, between income and crime (here income, which is itself a ratio, is being used as if it were a measurement or an actual object), between the numbers of books on mathematics and the intelligence of the population, are all cause-and-effect ratios. In the previous type of ratio we had a situation where both quantities were independent and the ratio defined the relationship between them. In cause-and-effect ratios, however, the assumption is that one quantity is dependent on the other and the purpose of the ratio is mainly to find the nature of this cause-and-effect relationship. Does an increase in smoking result in an increase of lung cancer? If so, is the increase in proportion or what? These are the kinds of questions which we try to solve by the use of cause-and-effect ratios.

The alert reader will have noticed that some, if not all, of the ratios in the last category belong to the realms of opinion rather than fact. What many people do not realize is that an opinion or an assumption, even when clothed in precise-looking mathematical jargon, is still only an opinion or an assumption.

In the early days of photography some enthusiast coined the phrase, "The camera cannot lie." That has been thoroughly debunked. We know that, even apart from deliberate faking of the negative, it is possible merely by taking a shot from a different angle to make a person look imposing or ridiculous, to make landscapes either beautiful or dreary, to make buildings appear magnificent or quite unstable and ill-proportioned. Mathematics, particularly where cause and effect ratios and graphs and statistics are concerned, can be every bit as bad as the camera. The following example shows how a graph can provide us with quite useful information or—if we take too much for granted—can become very misleading indeed. We will call it the case of the stretching rope.

Now even the idea of making a graph to give us a picture of the ratio between the pull on the rope and the stretch means that we are assuming that the stretch is the result of the pull. This may seem a reasonable and even obvious assumption but we should keep in mind the fact that it *is* an assumption. Having decided to make this assumption and to produce a graph we set out our two scales at right angles to each other. In the case of the conversion scales already mentioned it was necessary for equivalent markings to coincide. This meant that we could choose one scale but the other one then depended on the size of the first. With the graph, however, we have a free hand. The scales in no way depend on each other and we can make both whatever we please. Since our information is that the rope stretched two feet when a pull of 200 pounds was applied, it would be convenient to make the horizontal scale represent a stretch of—say—up to three feet (to give a margin for extending our graph) and 300 pounds on the vertical scale. We draw these scales and then find a point as before. We measure vertically up from the mark representing two feet stretch and horizontally across from the mark on the vertical scale representing 200 pounds pull and we find the point representing the information we have been given. Like this:

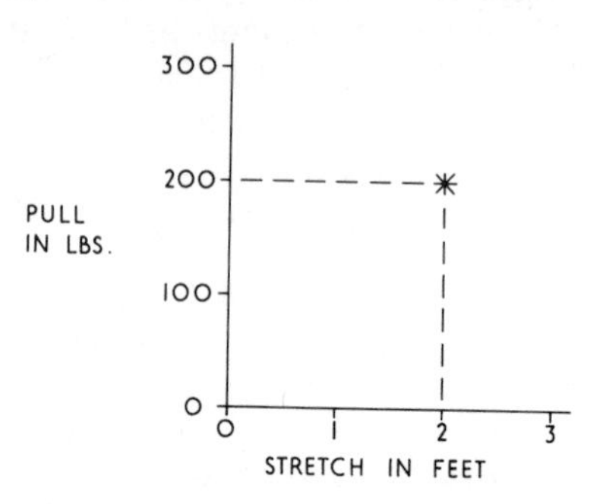

It would also appear obvious that when there wasn't any pull on the rope there wouldn't be any stretch either. So another point on the line representing the relationship between the pull and the stretch should be the intersection of the two scales where both are zero. Now if we make one more apparently quite reasonable assumption, that the stretch is proportional to the pull (for example, that the stretch would be one foot—which is half of two feet—for a pull of 100 pounds—which is half of 200 pounds—and so on) then we will find that all points representing stretch-pull relationships at different pulls will lie on a straight line joining the zero point at the intersection of the two scales to the 200-pounds-pull-2-feet-stretch point. (See sketch on next page.)

We could then say that this line represented the relationship between the pull and the stretch and we could, for any particular pull, find the appropriate stretch.

And the answers in spite of their apparent accuracy would be completely wrong!

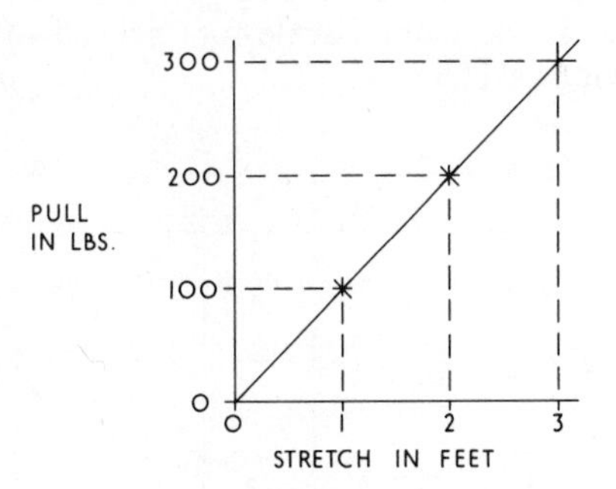

Up to the present there are only two basic facts we know about the stretch-pull relationship of the rope: first, that it had a certain length when it started with no pull (we must have measured this in some way to be able to measure the amount of stretch); second, that it stretched two feet from its original length with a pull of 200 pounds. If this pull is removed and we measure the rope again we will probably get a horrible shock to find that even though the pull has been removed the rope does not return to its original length.

This incidentally is what would be quite likely to happen in practice. The rope after being pulled would take an initial "set" because the pull causes the fibers to bed down and it never quite returns to its original length. So having found that not even our starting point is accurate it would be common sense to make some more checks at different pulls. We would probably get a table something like this:

PULL (POUNDS)	STRETCH (FEET AND INCHES)
0	Initial stretch 6 ins.
50	1 ft $1\frac{1}{2}$ ins.
100	1 ft 6 ins.
150	1 ft 9 ins.
200	2 ft

and our graph would be as is shown in the sketch below. Just to show that it doesn't matter which way the graph is drawn we have in this case used the vertical scale for the stretch and the horizontal scale for the pull—the opposite to the previous sketches.

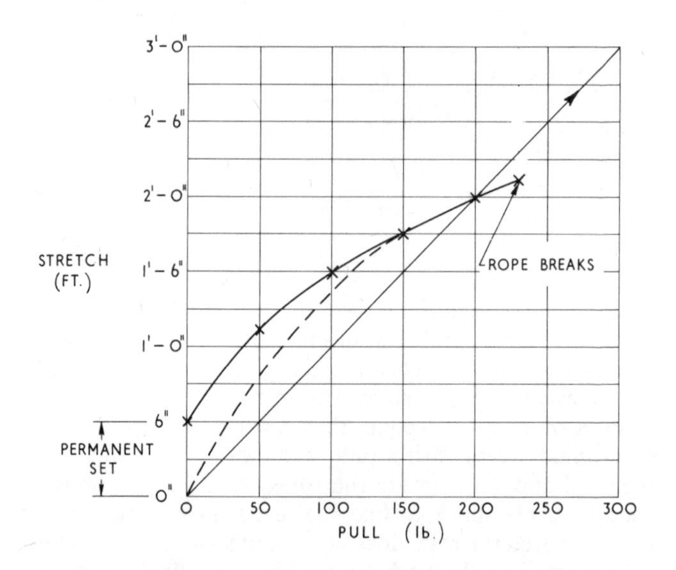

The heavy line represents the actual relationship between the pull and the stretch. The dotted line represents the relationship that would have existed before the rope gained its "set" and the thin straight line represents what we had first imagined the relationship would be. We can see that for any pull below 200 pounds we would have badly underestimated the stretch of the rope. Our estimate, for instance, was a stretch of 6 inches for a pull of 50 pounds and in reality the stretch is 1 foot $1\frac{1}{2}$ inches, more than twice that amount. Again the straight line would pass through a point representing a 3-foot stretch at 300-pound pull. Experiment shows that the rope breaks at 230 pounds and so the line cannot possibly continue beyond this point. We can see that when making a graph like this for practical use we have to be very careful what assumptions we make or the results can be disastrous.

On page 101 we showed how we could divide the pull by the stretch of the rope or vice versa and get an average value of the ratio. We can do this, and the straight line on the graph on page 110 does in fact represent this average value. Why it should be called the average value heaven only knows, because it hasn't the same accurate meaning that the word average has when we talk about the average height or weight of a number of people, or an average speed. In this case we see that right up to the 200-pound mark the actual stretch is in every case more than the "average" value. And, of course, if we had picked on the pull–stretch ratio at 100-pounds pull, we would have got a different "average" value. Perhaps the main use of the "average value" of the ratio in the example is to warn us not to put too much trust in mathematics and its definitions!

The case of the stretching rope was an example where the main danger lay in making unjustified assumptions about the details of the relationship between the pull on the rope and the resulting amount of stretch. But at least in this case we were justified in assuming the basic relationship existed.

Sometimes, however, we can't even assume this. Let us have a quick look at the results we get if we assume there is a relationship where none really exists at all.

Imagine that we have some statistical information about the population of a country town which came into existence in 1850. Before that time nobody lived in that area. We have the following information which is based on reliable records:

YEAR	POPULATION	DEATHS PER YEAR	NUMBER OF DOCTORS
1840	0	0	0
1850	100	0	0
1860	500	10	1
1870	1000	20	2
1880	2000	40	4
1890	3000	60	6

Now suppose we decided to make a graph giving us a picture of the relationship between the number of deaths per year and the number of doctors practicing in the town. It would look something like this:

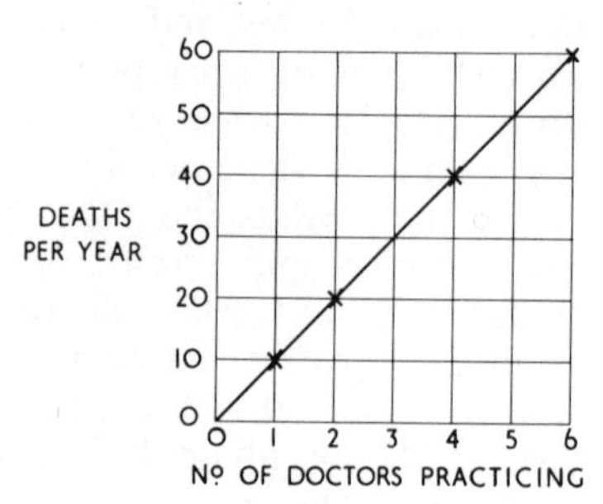

The crosses represent the points which we have plotted from the statistics, which, we repeat, are perfectly reliable. No doctors, no deaths per year; one doctor, ten deaths; two doctors, twenty deaths; and so on.

What does this graph tell us? It tells us with complete mathematical accuracy that for every doctor who comes to practice in the town, ten more people die every year!

Doctors may make mistakes, but nobody in his senses would produce a graph such as this as evidence of medical incompetency. Yet how is it that a graph, accurately plotted from reliable figures, can be completely misleading?

The answer is that we have assumed a relationship where none exists. There is no direct relationship between the two things we have compared. The number of deaths per year increased simply because there were more people living there. When there was nobody living there nobody died. The number of doctors practicing also increased because more people lived in the township.

This example has been deliberately exaggerated to the point of absurdity. But it is surprising how often we find graphs—often prepared by well-meaning and enthusiastic people—which are almost as unrealistic and equally misleading. Take for instance the controversial issue of lung cancer. This is increasing, so is cigarette smoking, so is man-made air pollution and so also, by the way, are old-age pensions. It is perfectly easy to get accurate figures on all these things and to draw graphs which show that lung cancer is increasing with increasing smoking, or with increasing air pollution, or even with increasing old-age pensions. When faced with such graphs people often question the figures, and are surprised when these turn out to be reliable. What they should question is whether any valid relationship exists between the two things being compared. And this is one question which mathematics cannot answer!

The fourth and last type of graph we are going to mention is what we could call the record type of graph. It is still a ratio, but usually the relationship is between some variable factor and the passing of time. For instance, we could make a picture of the number of people in a town over a period of years and see if it was rising, falling, or fluctuating. We can draw a graph which gives a picture of a person's income, or a business firm's profits over a period, and so on.

At first sight it might seem that such graphs would be delightfully simple and foolproof. They are accurately plotted from recorded figures, there is no assumption or guesswork. They are usually plotted against time in years or hours or some other convenient unit, and there is no danger of assuming some non-existent relationship.

In actual fact, however, unless we know exactly what to look for, these "simple" graphs can be more misleading than all the others put together. And because they are so important and can be so misleading they have been given a little chapter all to themselves.

6. ...AND RACKETS

There are lies, damned lies,
and statistics.

Attributed to Disraeli

The main purpose of what we have called the record type of graph is to give a picture of some change; a picture that we would not see clearly if we only had tables of figures.

All of us have seen—in cartoons or in reality—the temperature charts which are kept in hospitals as a record of the fluctuations in the patient's temperature. If it were only necessary to know what the actual temperature was at any particular time there would be no need for a graph; a table of figures would do equally well. Look at the example below. It is a table recording the temperature at hourly intervals.

Time	Temp.	Time	Temp.
1 p.m.	97·5	7 p.m.	102
2 p.m.	97·7	8 p.m.	103
3 p.m.	97·9	9 p.m.	100
4 p.m.	97·5	10 p.m.	99
5 p.m.	97·9	11 p.m.	98
6 p.m.	99		

We can see from the figures that the temperature went up and then down again, but how much more clearly this can be shown on a graph is illustrated by the one at the top of page 115.

This gives us a much better picture of the variations of temperature than we could ever get from looking at a table of figures. We see that at about 5 p.m. the temperature started to rise sharply and that after 8 it fell equally sharply.

But while the graph is capable of giving this quick and clear picture there are a number of ways in which it can prove most misleading, precisely because it appears so clear and unequivocal.

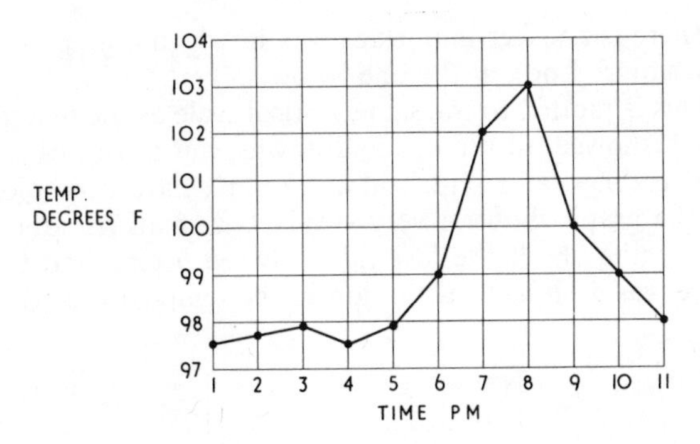

If you were a patient in the hospital—and assuming that the absence of large fluctuations in temperature meant you were getting better—which of the two temperature graphs would you prefer, the one above or the one shown in the graph below?

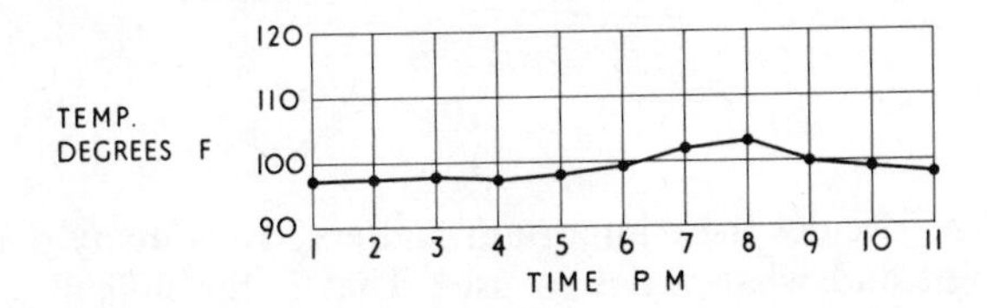

Most of us would immediately settle for the second graph because the line is much more level. But if we look at the two graphs carefully we will see that they are exactly the same—in fact they have both been plotted from the figures given on page 114. They both tell exactly the same story but the vertical scale has been changed. In the first graph we can see that one division represents one degree change in temperature, while in the second the same vertical distance represents not one but ten degrees change in temperature. So the ups and downs of the graph are compressed to one-tenth of what they were in the previous figure. They might *look* better but they represent exactly the same changes as are shown on the other graph.

There is another important way in which a graph can be misleading. Look at the one below.

This is plotted to the same vertical scale as the first graph which showed all the ups and downs, but in this case only two readings, at 1 p.m. and at 11 p.m., have been plotted. So the graph shows a very steady but small rise and tells us nothing about the fact that between 6 p.m. and 9 p.m. there was a violent fluctuation in the temperature.

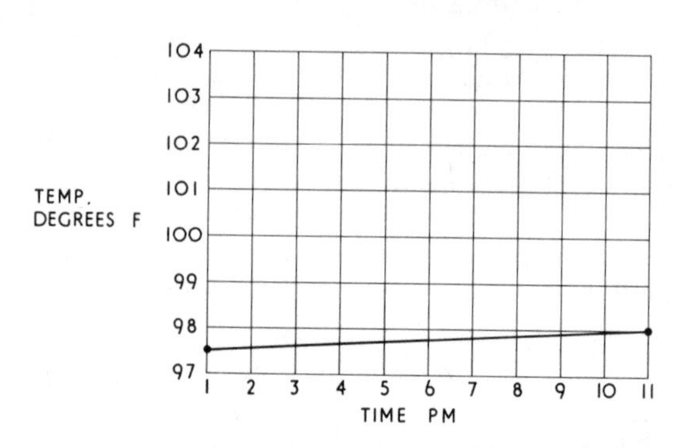

Who decides these important matters? How many points to plot and what scales to use? That is the difficulty. In practice the person preparing the graph selects a scale, and uses the amount of detail which he considers will produce a pictorial impression of what he wants to convey. The trouble is that most of us feel that a graph, provided it is plotted with mathematical correctness, has some objective reality regardless of the ideas of the person who prepared it. To some extent this is true but to a far greater extent the graph simply expresses not a mathematical truth but the personal opinion of the person who produced it.

Without being too uncharitable it could be suggested that nowadays there is a tendency in company reports, political propaganda, sales propaganda, and similar fields to develop a "simple graph for simple people" technique. In the bedtime story in graphs that follow we will see just how far this kind of thing could go. All the graphs are perfectly "genuine" in the sense that they have all been plotted accurately and are all from the same table of figures. Finally, of course, it must be emphasized that all characters and factories and so on in this little story are fictitious and that none of the people we know would ever dream of misleading their fellow creatures.

The story begins with a rather worried company accountant. He was faced with the problem of preparing two reports. One was for the approaching shareholders' meeting to prove to them that production was increasing by leaps and bounds and that they couldn't possibly invest in a firm with a brighter future. The other report was for an arbitration conference with the employees, and this had to prove that production was increasing so slowly that the firm couldn't possibly afford to pay increased wages.

The actual production figures which, as he was a scrupulously honest accountant, he couldn't dream of faking, are shown in the table below.

YEAR	NUMBER OF JARS PRODUCED
Start	100,000
1	110,000
2	120,000
3	130,000
4	140,000
5	150,000

The first graph which the accountant drew up looked like this:

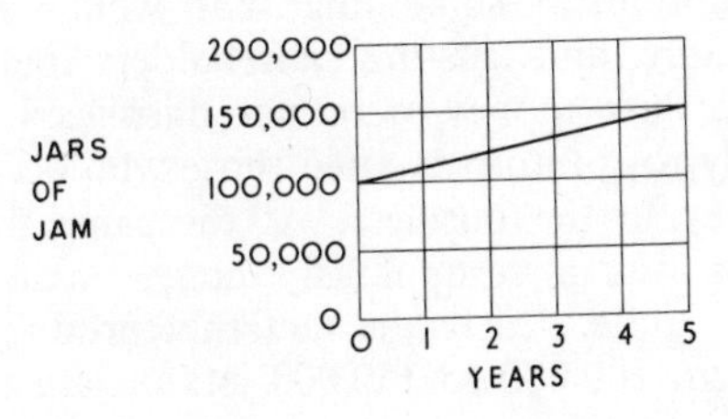

As far as the shareholders were concerned it was not really inspiring enough. The union, on the other hand, was sure to claim that the rise in the production justified a wage rise. After hours of contemplation and a wastepaper basket full of graphs, the accountant produced a graph for the union conference and it looked like this:

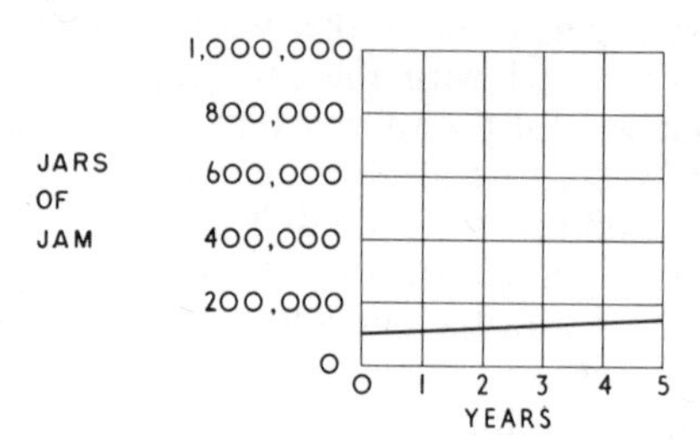

The accountant was very pleased with his effort. He had an artistic soul and was concerned with the True Meaning of Things and not with Mere Mechanical Facts which could be misinterpreted by an ignorant observer. He felt that the graph put over the message that production was only increasing at a snail's pace compared to what it might have done, and the firm couldn't possibly afford to pay more wages. At the same time it gave the exact facts, that production had increased from 100,000 to 150,000 jars of jam in five years.

The shareholders' meeting was a very different proposition. The accountant's sensitive artistic soul told him that in this case what was needed was a note of inspiring and cheering optimism—something that would convince the simple, kindly, and trusting shareholders that their confidence and their money were not misplaced; and that a particularly rosy future awaited those who decided to keep their money in the business. At the same time—for the accountant was a scrupulously honest man—the graph must show quite clearly the fact that production had increased from 100,000 to 150,000 jars of jam in five years.

The graph which finally satisfied the accountant's inner artistic convictions looked like this:

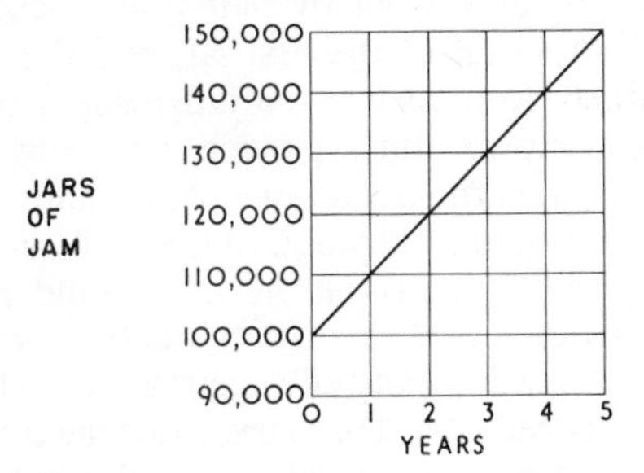

But on thinking it over he began to wonder if he could not do better still. The graph was perfect from a technical and artistic point of view but the shareholders needed something warmer and more human—perhaps a picture of a jar of jam. Thus the accountant got his inspiration. He would show them not a plain graph but a number of jars of jam, one for each year, and the height of the jar would correspond to the production level for that year. So, using the vertical scale he had already worked out for his graph, he made a picture like this:

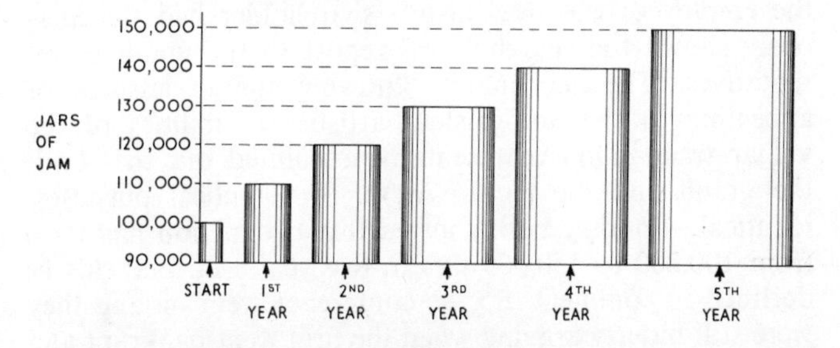

The accountant was delighted with the result. It was at once both homely and inspiring. The proportions of each jar were completely accurate and the vertical scale on the left clearly and unmistakably showed—for those who were interested—the actual production figures for each year.

The only jarring note in the chorus of praise was struck by a nasty little office boy who pointed out that though the production had only risen by half, the height of the last jar was six times that of the first jar, and the area—which was more likely to influence the judgment than the height alone—was thirty-six times greater, and thus an increase 24 times more than the actual one. And finally, the amount the largest jar would hold was 216 times the amount in the first jar. So the contents of the jars would represent an increase 144 times greater than the actual production increase and heaven help any silly shareholder who looked at the pictures instead of at the figures. But the office boy only said all this to the junior typist and she knew he was trying to show off the knowledge he had gained at night school, so nobody took any notice. Soon afterwards the office boy resigned to begin selling newspapers and become a self-made millionaire.

The shareholders' meeting was a complete success and, largely as a result of the comforting picture, the shareholders never had a moment of anxiety about their money right up until the firm went bankrupt a few months later.

The result of the union conference however was not so happy. The accountant duly produced his chart but one of the employees who was also a shareholder had treacherously shown the shareholders' report to the union representatives. The accountant, knowing the uselessness of appealing to the non-existent artistic sensibilities of the vulgar trade-union representatives, pointed out that both the graph and the picture were, for practical purposes, identical—for they both showed that production had risen from 100,000 to 150,000 jars in five years. Beyond this he declined to comment. So the conference went on and they were still bitterly arguing when the firm went bankrupt and the dispute became of academic interest only.

A little before this sad event, however, the manager of the factory found himself in a rather awkward situation. In spite of the successful shareholders' meeting the directors had an uneasy feeling that something was wrong. So they naturally decided to investigate managerial efficiency at all levels (below the board of directors) and the possibility of cutting down staff. The manager was ordered to produce a graph showing the relationship between the production and the staff employed over a period of five years.

At first the manager felt very pleased at this opportunity to show his efficiency. Since he had taken over there had been an ever-increasing number of people rushing energetically around. He collected the figures showing numbers of staff from the factory records, got the production figures from the accountant, and then made up the following table:

YEAR	NO. OF STAFF EMPLOYED	NO. OF JARS OF JAM
Start	100	100,000
1	140	110,000
2	180	120,000
3	220	130,000
4	260	140,000
5	300	150,000

As he looked at the table the manager's face grew longer and longer. When he had taken over there had been 100 people producing 100,000 jars of jam a year. Now there were 300 people producing only 150,000 jars a year, which made an average of 150,000/300 or 500 jars per person where previously the figure had been 1,000 jars per person. The manager frantically re-checked his figures but they remained the same. He racked his brains to find the explanation. He knew it couldn't be inefficiency because he had been in charge of everything himself. Then he realized the explanation must lie in the increase in the quality of the jam. Obviously with three people stirring where one had stirred before and three people inspecting where one had inspected before there must be an increase in quality. He telephoned the chairman of directors and suggested that the board might like some graphs showing the increased quality of the jam compared to the numbers of staff employed. The directors, however, were most pig-headed. All they wanted was a comparison between the staff and the production.

So the manager plotted a trial double graph using two vertical scales, one to indicate the jars of jam produced and the other to show the figures for the staff. The finished graph looked like this:

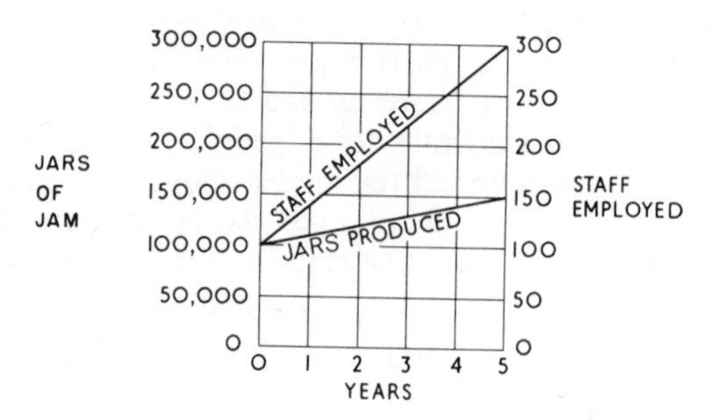

It was clear to the manager that this graph, though it might give a superficially and mechanically accurate picture, could not possibly represent the basic reality of the situation. It could even be interpreted by uncharitable directors to indicate inefficient management, which was of course impossible. So he had another try. He used the accountant's production graph and made up one of his own for the staff. They looked like this:

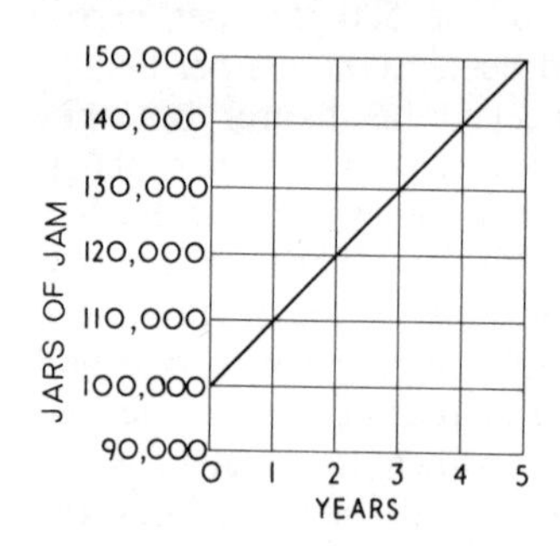

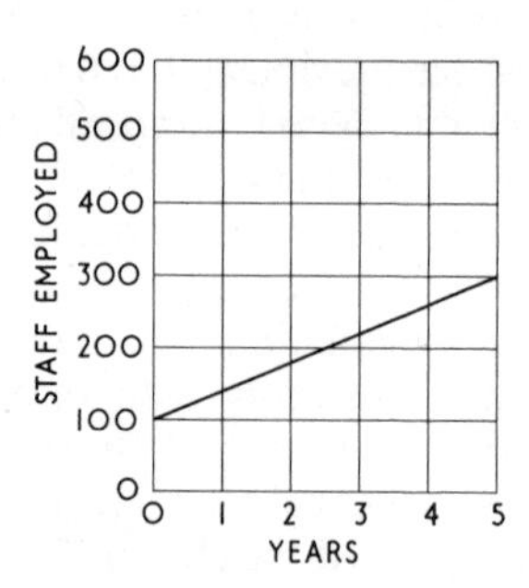

The manager felt much happier about these two graphs and sent them on to the board of directors. Unfortunately there was one director with a most unpleasantly suspicious nature. He wanted to know why both graphs had not been combined into one, and why they did not both start at zero. He complained that it was impossible to get a real comparison between them. So they were returned to the manager.

But the manager was ready, and produced a masterpiece. He combined them both into a single graph and what was more the vertical scale started from zero and the one scale did for both the staff and the jars of jam. And the manager's final graph looked like this:

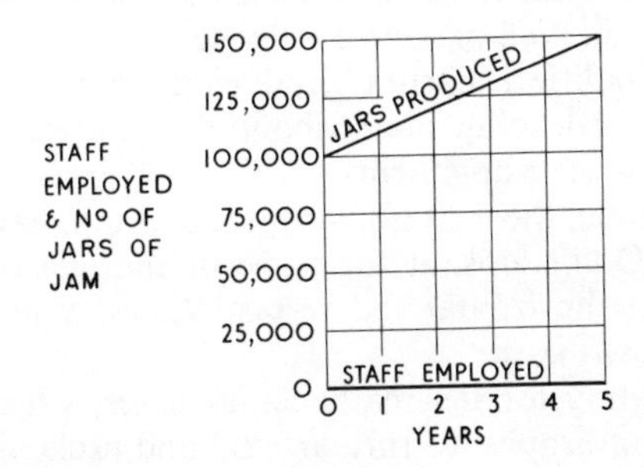

The line showing the change in number of staff was not exactly revealing. As one division represented 25,000 units and the staff numbers began at 100, the commencing point was 100/25,000 or 1/250 of a division from the base line and the finishing point was 300/25,000 or about 1/83 of a division, which is about the width of a very thin pencil line.

The manager sent the graph along to the directors, pointing out that he had done exactly as requested and he was sure the new graph would give them all a perfectly clear picture of the situation. If anything it told the board of directors slightly less than the first graphs did. But they had got exactly what they had asked for and no one was prepared to admit he was so ignorant that the graph meant nothing to him. So each director looked at his copy with an air of profound wisdom and said, "Ah yes, I see," several times; and then the chairman asked the secretary to introduce the next item on the agenda. And so we leave them there in an atmosphere of calm and dignified contemplation and come to the end of our bedtime story. If the story has given some indication of how the same figures can be plotted to give lines which rise sharply, or rise gradually, or don't rise at all, and how fast or slowly rising figures can be made to look the opposite, it will have served its purpose.

In real life, of course, we *never* find in a company prospectus a graph which makes production figures look better than they are. We never find managers producing graphs which are more likely to bamboozle their superiors than to enlighten them. And of course we never, never find pressure groups or political parties producing graphs or charts which give misleading ideas about the relative merits of themselves or their opponents.

Just the same, the next time you see a graph, have a good look at it. Don't look at the slope of the line but at the figures on the horizontal and vertical scales. You may find the results instructive.

It is tempting for the mathematics hater, when he sees these kinds of graphs, to turn around and exclaim, "That's mathematics for you!" To do this, however, would be to make of mathematics a scapegoat for the iniquities of man. Mathematics is not an agency for good or for evil, but purely and simply a tool which all kinds of different people can use to get all kinds of different results. The fact that there may be a few mathematical spoilsports around is hardly a reason for condemning the tool which they use (or misuse).

We will finish on an appropriate note by quoting a rather sad little piece of verse—sad from its writer's point of view anyway—addressed to the rapidly departing figure of a bookmaker (the racetrack type, not the literary one).

> What matters that I backed a noble steed,
> Which started well and all the field outran?
> I did not reckon, bookie, on *thy* speed.
> The proper study of mankind is man!

In the long run, perhaps the best way of checking the validity or otherwise of the kind of graphs which have been illustrated is to find out something about the kind of people who produced them!

EXERCISE

1. Make up a graph which will convince your employer that, taking both the increases in your salary and the increases in the cost of living into account, you are far worse off than you were five years ago.

7. THE RATE OF CHANGE

What's not destroy'd by Time's devouring hand?
Where's Troy, and where's the Maypole in the Strand?
Bramston

In the previous section we discussed graphs and how they could be used (or misused) to give a clear and simple picture of the way in which one quantity varied in relation to another. A number of these pictures could then be compared, but the main problem was to make sure that the comparisons were not misleading.

In this section we will continue with graphs, taking a closer look at some of the fundamental ideas concerning change and showing how these ideas can be pictured in the form of graphs. This will help to give a clear conception of what we mean when we talk about change and rates of change.

Everyone of course knows what change is. It's simply—change. One dictionary gives the meaning as "to alter." When we look up the word "alter" we find it means "to effect a change in." All of which is very clear and obvious if not particularly helpful.

If, however, we do make a real attempt to explain what change is we will perhaps come to the conclusion that it is something which happens over a period of time. Then we have to try to explain what time is, which makes things more difficult still. But if we persist with the problem for long enough we may very well come to the conclusion that the idea of time and the idea of change are practically inseparable.

It is impossible to think of change without having in our minds a kind of "before" and "after" picture which of course brings in the idea of time. In the same way it is impossible to think of the idea of time without bringing in the idea of change. In fact the only way we have of measuring time is through the medium of some kind of change—usually the change in the position of the hands of a clock. It is no accident that the word "timeless" is such a fitting description of vast desert areas or mountain ranges where it is impossible even over long periods for us to see any change in the landscape. Perhaps you may remember sitting beside a still lake on some drowsy summer afternoon and experiencing this sensation of timelessness . . . , the feeling that everything is "standing still," in other words, that nothing is changing.

What then is time? Is it the devouring monster described by the Reverend Bramston? Or is it an ever-rolling stream bearing everything away to some other place? Or could time be likened to the irascible old gentleman who revenged himself on the Mad Hatter by keeping the time at six o'clock tea time all day and every day for months? In fact, time is none of these things. It doesn't even exist, except in our imaginations!

In the previous chapter we saw that in the mathematics dealing with change and comparison there was a kind of division with a difference called a ratio. We also saw that some of these ratios, such as speed, which is a comparison between distance and time, and pressure, which is a comparison between a force and an area, are so important that we have come to look on them as things which have an existence of their own and not merely the result of a comparison. You may remember that way back in Chapter 1 we saw how easy it was to get into a mental muddle if we confused ideas with things. We can put a rope around a tree stump and pull it out of the ground, but we would have some difficulty in trying to remove a hole by the same method.

If we have a closer look at some ratios we will see that the same distinction applies. Take income, for example. If we are fortunate enough to have some money, we can put it in our pocket and go shopping. We can put it in a bank and get a receipt, which is really only another form of money. If we buy something, we hand the money over to the shop cashier and when it is spent we haven't got it any more. Before you decide that this is all too obvious to be worth mentioning let us compare money with income. If we spend all our money we will have no money left, but we will still have exactly the same income as we did before. We can't spend our income, we can only spend the money which we get as a result of having an income. We can't see or feel our income, and, although it may seem very real and is certainly useful as an idea, it is something that exists only in our minds.

The same thing applies to speed. We can see a real car moving along. We can see the road it is travelling on. Experience gives us a good idea of what its speed will be. But we can't see the speed, or hear it, or put it in a brown paper bag and take it along to the court as evidence if we are accused of dangerous driving. And exactly the same kind of situation arises when we begin to think about time. It is easy to imagine time as being a kind of invisible river which gradually erodes everything as it passes. It is better to have some picture than no picture at all, but if we begin to think of time as being something tangible like the waters of a river we can arrive at a mental attitude which limits our understanding of reality. In the opinion of the writer a clearer and more coherent viewpoint can be based on the idea that everything in the universe is in a state of change and movement, and that our conceptions of time arise from our need to set up some standard—having, as far as we can see, a constant change—against which we can compare all other changes. Just as we would be in complete confusion if we tried to measure things with an elastic ruler, so we would be in the same kind of predicament if we tried to measure the time of day with a clock which continually varied the amount by which it was changing.

An interesting sidelight on this is the fact that the earliest makers of ships' chronometers spent their energies trying to make instruments which were not so much accurate as constant. A chronometer which gained five seconds one day and lost perhaps ten the next would be quite unpredictable and useless on a long voyage. But a chronometer which gained even as much as fifty seconds a day—as long as it gained exactly the same amount every day—would be perfectly satisfactory. All one had to do was to subtract fifty seconds for every day since the chronometer had been set, and one knew the exact time.

But in the last resort we can say only that clocks keep exact time because they change at the same rate as we do. Our heart-beats, on an average, have the same kind of regularity. So does our breathing and so do our brainwaves. But if we were like plants and insects, which speed up during some seasons and slow down during others, we would—if this same effect happened to everyone together—be convinced that the clocks and the days and everything else around us was speeding up and slowing down. No matter how hard we try we cannot know anything beyond our senses and we cannot know of any "standard time" which may exist outside our experience. So in the last resort all our ideas of time must be based on the fact that we are changing.

Some materials, such as certain metal alloys and the quartz crystals used in modern electronic time standards, undergo a process called "aging." This is essentially a change in some of their qualities. The snag is that the rate of these changes varies and we can have an old crystal which is "young" and a young crystal which is "old." In these cases the only satisfactory classification of the "age" is by the amount of change that has taken place. The length of time the material has been in use is meaningless in assessing its "age."

In the universe there are, of course, many things which are changing so slowly or so slightly that by ordinary standards they don't seem to be changing at all. But if we check far enough back we find that even these things join in the universal process of change. From the planets and the sun down to the atoms, everything is changing its position, or its nature, or both.

It was only after the beginning of the seventeenth century

that mathematics really began to be applied to problems concerning comparison and change. We have already seen how, in order to cope with these new ideas, the ratio, a queer kind of division which wasn't division, had to be introduced into mathematics. And from this came graphs and conversion scales and all the ideas and methods existing today. We have already seen, however, that the rate of change as shown on a graph can be misleading if compared to another rate of change shown on another graph which has different scales. So one of the first and most important things to learn is how we can find some standard way of measuring rates of change which will give consistent answers regardless of how the graph happens to be drawn.

All motorists and cyclists are familiar with hills. When a hill has a small change in height in relation to the distance travelled we say it is a slight hill. But when the change in height is large in relation to the distance travelled we say it is a steep hill. A very simple way has been worked out for measuring the amount of steepness (or more generally the rate of change of height per unit length) and it is shown in the sketch below. What we do is simply examine the amount which we have travelled upwards and divide it by the amount we have travelled along, like this:

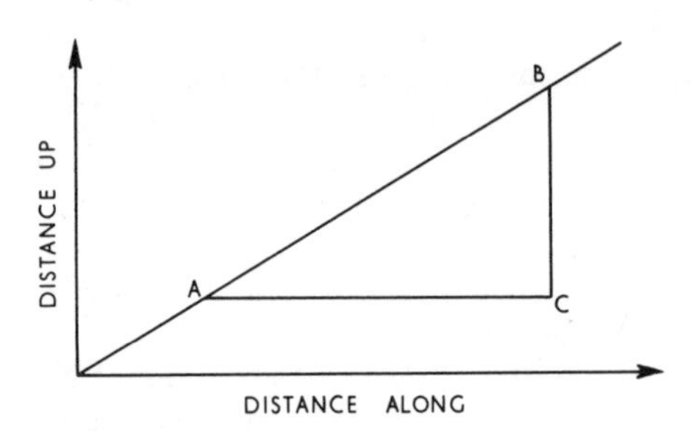

In the figure above the distance from C to B represents the height and the distance from A to C represents the length along when we travel from point A to point B.

So the slope is equal to the length CB divided by length AC,

$$\frac{(\text{LENGTH } CB)}{(\text{LENGTH } AC)}$$

It is worth remembering that although we have an apparently simple division of one length by another giving a simple single answer we are really dealing with a ratio. If for instance the slope is quite steep and CB is 4 feet while AC is only 2 feet we have the slope equal to CB/AC equals 4/2, equals 2. This doesn't mean two inches or two feet or even two "slopes." It means a slope which goes up two units for every one unit along. We should always think of a slope being two (up per one unit along) or three (up per one unit along), or, if it is not so steep, as a fraction like one quarter (up per one unit along) and so forth.

Even if these ideas only applied to real slopes, such as we have on hills and on the roofs of houses, they would be quite useful. But we can do much better than this and apply the same idea to any ratio we can think of. We have already seen how we can make a "picture" of things like time, force, pressure, or speed by representing units of these things as

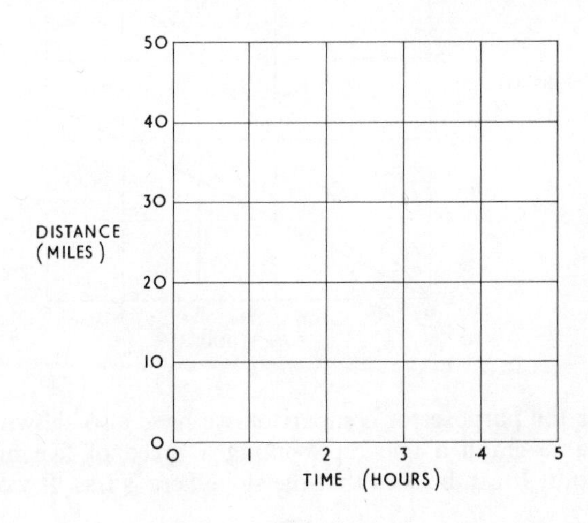

units of length along a line. Suppose, for example, we wanted to make a picture representing a speed of ten miles per hour. Since speed is a ratio between an amount of distance divided by an amount of time we would have to draw a vertical scale representing distance in miles and a horizontal scale representing time in hours (shown above).

If we are travelling at a speed of ten miles per hour it is clear that in one hour we will have travelled a distance of ten miles. In two hours we will have travelled twenty miles and in half an hour we will have travelled five miles. Therefore if we are to draw a line which will represent a "picture" of a speed of ten miles per hour it must pass through a point which is five miles "up" and half an hour "along," and another point which is ten miles "up" and one hour "along," and a third point which is twenty miles "up" and two hours "along" and so forth. The line going through all these points and all other similar ones would look like this:

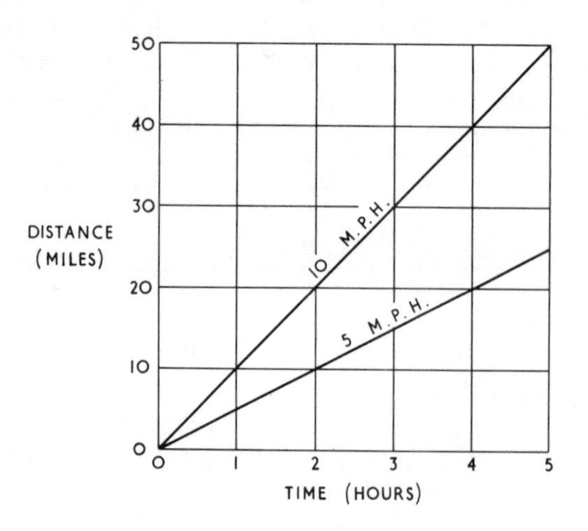

For the purposes of comparison we have also shown on the same graph a line representing a speed of five miles per hour. It can be seen that the slope here is less. It works out that the higher the speed the steeper the slope of the line. In fact, the slope *is* the speed itself, and we will find that whatever kind of graph we draw and whatever the subject is, the slope at any place on the graph will represent the actual value of the ratio at that particular place. If we go ten miles up for one hour along, the slope will represent a speed of ten miles per hour. If we go 100 pounds up for a stretch of one foot along, the slope would be 100 and we say the rope requires a pull of 100 pounds per foot stretch.

If we plotted the graph the opposite way, with the stretch on the vertical axis and the pull on the horizontal one, we would get a slope of $\frac{1}{100}$ and we would say that the rope stretches $\frac{1}{100}$ of a foot per pound pull. You will remember that in Chapter 5 we explained how both these ratios meant exactly the same thing.

All this sounds quite simple, and so it is really. But there is a trap into which the unwary can fall. Let us look again at the previous figure. Looking at the graph squares we see that the 10 miles per hour line passes through the intersection of the first division up and the first division across, the second division up and the second division across, and so on. In other words it is going up at an angle of 45 degrees, half way between zero degrees which is the horizontal and 90 degrees which is the vertical. So the line shown has a slope of one in one or simply one. But if it represents a speed of 10 miles per hour it obviously should have a slope of 10 instead of a slope of 1.

The explanation lies in the fact that this graph, like many of the graphs in the previous chapter, is drawn in a thoroughly misleading way. The fact that there is a good reason for doing this doesn't make it any less misleading. If we had wanted to get a true picture of what the angle of slope actually looked like, we would have to make one unit on the vertical scale exactly the same length as one unit on the horizontal scale. If we did so our graph would look like this:

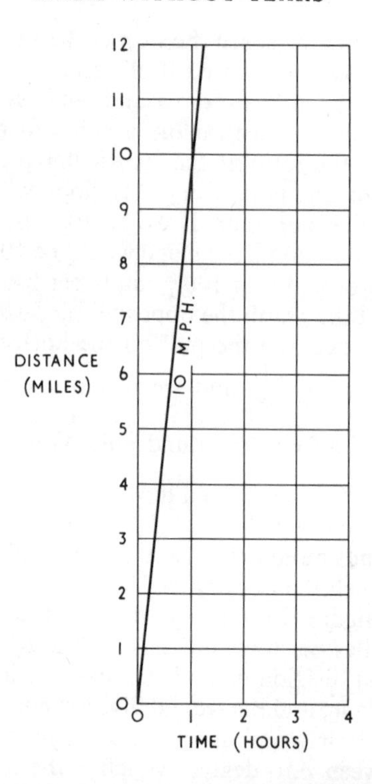

Although it has apparently no resemblance to the graph on page 132 it is in fact exactly the same line drawn this time in the right proportions, and we can now see quite clearly that the slope *is* 10, that is 10 (miles up) per 1 (hour along). Why aren't graphs always drawn to the correct scale? The answer is simply that in many cases we would get completely unworkable shapes. On the facing page is a graph with a line representing a speed of eighty miles per hour, over a time from zero to five hours. The distance scale must go to 400 miles which is the distance travelled in five hours if one goes at eighty miles per hour.

If we just looked at the graph we would guess that the slope as drawn was not quite one in one (actually it is four up for five along, giving a scale speed of $\frac{4}{5}$ miles per hour), and the angle at "*a*" is about 39 degrees. But if we drew the graph to the true scale we would find that the correct slope is 80 units up for one unit across and the angle at "*a*" is about 89½ degrees—so near to the vertical that it would be difficult to separate them.

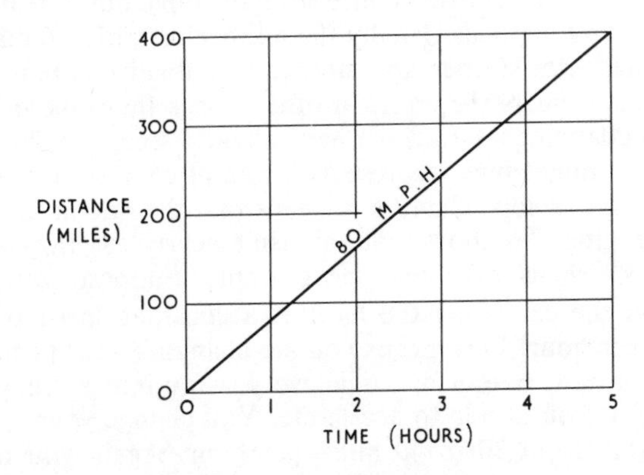

If you still believe that graphs should always be drawn so that the angle and the slope are exactly true to scale, then try drawing this graph above in its correct proportions and see what it looks like. You won't be stuck for something to draw it on. If you think a little you will realize you have exactly the right-shaped paper quite handy. Throughout the remainder of this chapter, when we refer to the slope, or the angle which it makes with the horizontal, we will be talking about the correct slope and not the distorted one which will exist on an illustrative graph when the lengths of the units on the horizontal and vertical scales are not equal. There are of course mathematicians who say that anyone who tries to estimate the nature of a ratio from looking at the slope of a graph deserves what he gets. Such people contend that one should only work with pure and uncontaminated mathematical symbols where calculations and not pictures are the proper thing. This may be all very well but it is like saying that the proper way to prevent indigestion is to eat only the plainest and most unpleasant foods which are known beyond doubt to be easily digestible. This book, however, has been written for human beings—not for mathematicians—and we will stick to our pictures.

Let us, however, get back to the hills we mentioned. The picture we drew showed a nice straight slope, but in real life this doesn't happen. Usually the hill begins fairly gradually and then gets steeper and steeper and finally flattens out again as it nears the crest. In other words the slope of the hill is changing. But, as we have already seen, the slope—even a straight one—represents a rate of change of height. So when the slope changes it means that the rate of change is changing. To show that this isn't nearly as crazy as it sounds, let us take another example. Suppose you are driving the car along the level at a constant speed of 30 miles per hour. This means you are changing your position at a constant rate of one mile every two minutes. Suppose now that you decide to accelerate. You change your speed smoothly from 30 to 40 miles per hour. Again you have been causing a change in the rate of change. But suppose you had started to accelerate from 30 miles per hour with your foot hard down on the accelerator and gradually eased off as you approached 40 miles per hour. Then you would have caused a change in the change of the rate of change, and that is really something! It's much easier to do it than to think about it!

But if you find this all a little hard to follow and you have to think twice over it, you may find consolation in the fact that the most brilliant thinkers and mathematicians of ancient times were completely unable to comprehend even the simplest—to us—ideas about the simplest kinds of ratios like speed and pressure. All the same, before we get too pleased with our superior intelligence, we should realize that we have been brought up in a world where speed is something we come into practical contact with every day, and acceleration is quoted in every car advertisement and road test report. It is because of our practical experience, not because we are more intelligent, that we can understand quite easily things and ideas which our ancestors could not even dream about. On the debit side we have problems they never even dreamed about. Like air and water pollution, for example. . . .

Getting back to change and rates of change and changing rates of change, let us see what some of these things look like when we try to make pictures of them in the form of graphs. We will begin with a graph representing something standing still, that is, something which has no speed at all. For this graph we will have a vertical scale representing distance and a horizontal scale representing time. The graph would look like this:

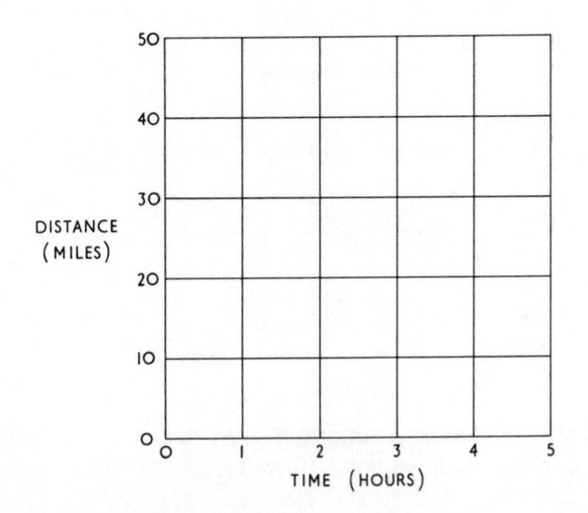

It's quite all right. We haven't forgotten to draw the line in. The fact is that the graph line representing zero speed is the same line as the horizontal base line. We start at zero distance and zero time. After 1 hour we still have travelled zero distance. So we have 0 miles for 1 hour, 0 miles for 2 hours, 0 miles for 3 hours, and so on. The line, however, doesn't have to be on the base line. Suppose we were talking about a stationary object which was 10 miles away. At zero hours its distance would be 10 miles, 1 hour later its distance would be 10 miles, and 2 hours later its distance would still be 10 miles. So in this case we would have a horizontal line at the 10 mile level. We see that in this case a horizontal line means no change of distance over a period of time—in other words, zero speed.

Suppose now that we are going at a speed of 10 miles per hour. The graph which we draw would look like this:

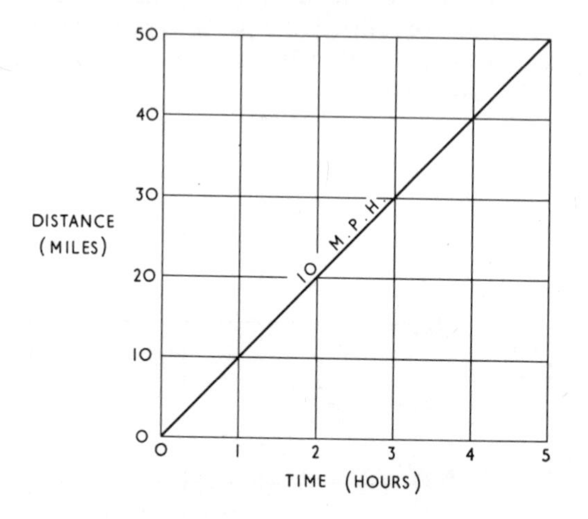

Starting from zero time and zero distance we find that after 1 hour we have gone 10 miles, after 2 hours we have gone 20 miles, and so on. Here the slope of the line (the correct slope we would see if the vertical and horizontal units were equal) is of course 10.

We don't always have to use distance and time for the vertical and horizontal scales of our graph. While we are talking about speed, which is the ratio between distance and time, it is of course the obvious choice. Speed is the rate of change of distance over a given time. But if we want to have a look at changes of speed instead of changes of distance there is no reason why we shouldn't use speed for our vertical scale instead of distance.

So let us take our steady speed of 10 miles per hour and show it on the new graph. It would look like this:

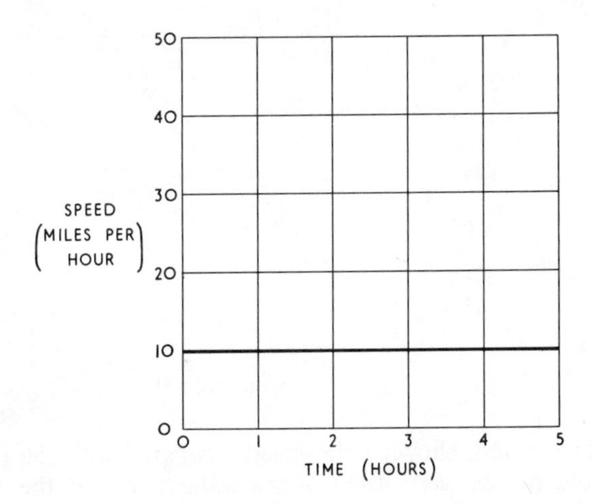

We are assuming a flying start, that is, right at the beginning we were travelling along at a steady 10 miles per hour. After 1 hour we are still travelling along at 10 miles per hour, after 2 hours we are still travelling at 10 miles per hour, and so on. Once again we are back to the horizontal line indicating no change. But this time it is not no change in distance but no change in speed. We have got past the first level of simple rate of change and are now talking about changes in the change.

Let us suppose that the speed is changing. Suppose we begin from a stationary position and by very patient and careful driving we organize things so that at the end of 1 hour we have reached a speed of 10 miles per hour, and at the end of the second hour we have reached a speed of 20 miles per hour, and we keep on increasing our speed (that is, accelerating) at the rate of 10 miles per hour every hour. Again we can have a vertical scale representing speed, and our graph will look like this:

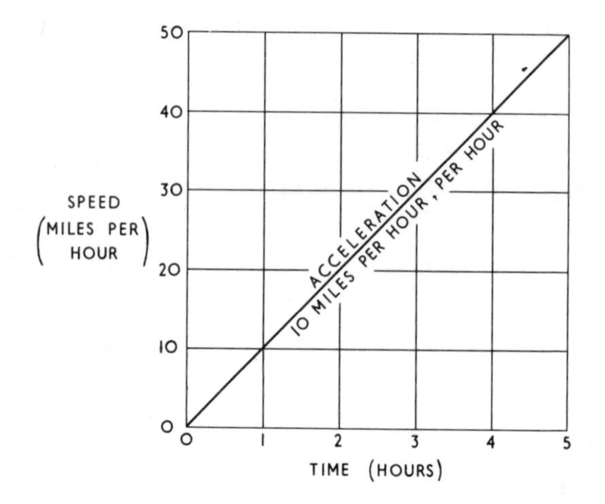

This graph, showing the speed compared with the time, would not be very useful if we wanted to find the total distance travelled at any particular period. To get this it would be better to make a graph comparing time and distance as we did when we were discussing simple speed.

When we come to do this we find we have to be careful. When we plotted the graph of simple speed on page 132, we could very easily find the distance travelled during various periods at a speed of 10 miles per hour. It was 10 miles after 1 hour, 20 miles after 2 hours, 30 miles after 3 hours, and so on. But with the graph of acceleration shown above, we start from a stationary position and it takes a whole hour to reach a speed of 10 miles per hour. Obviously during most of this first hour the speed would have been well under 10 miles per hour and thus the distance travelled in this first hour would be well under 10 miles. Actually, if we tried it, we would find it was 5 miles. During the next hour we would be travelling at between 10 and 20 miles per hour and the total distance would increase to 20 miles. From then on travelling at between 20 and 30 miles per hour we would have done a total of 45 miles at the end of the third hour.

Having found a number of these distances, we can plot these points on a graph, like this:

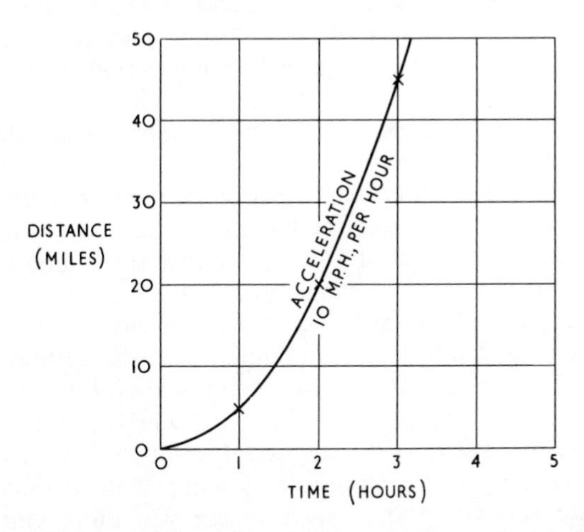

If we draw a line through these points we find it is not straight but curved. If we think about it we will see that it must be so for a graph of distance plotted against time is a graph in which, as we saw back on page 133, the slope represents speed. And since with acceleration the speed is continually changing it follows that the slope must be continually. changing. But in spite of the fact that the line is curved the slope of the line at any particular place still represents the speed at that particular place, just the same as when we had a straight line.

So we see that if we consider acceleration as a change in speed and make a comparison between speed and time we get a simple straight line graph. But if on the other hand we consider acceleration as a change in the rate of change of distance, and make a comparision between distance and time, we get a more complicated line which is curved. This provides us with a quick way of telling at a glance whether we are dealing with a simple or a more complex relationship.

Before we really start going around in circles, it might be as well to have a brief look at where we have come from and where we are going. We began with the simple idea of distance. Then we had the idea of speed, which is a change of distance over a period of time. Then we came to acceleration, which is a change of speed over a period of time or, to put it differently, a change in the rate of change of distance over a period of time. There is no reason why we should stop here. We can go on to some unnamed idea which is a change of acceleration over a period of time or, putting it the other way, a change in the change in the change of distance over a period of time. If you feel in the mood, you could plot a graph showing, for want of a better word, "thingamabobs" in miles per hour, per hour, per hour, with our old stand-by, time in hours, along the bottom, and distance as our vertical scale. One could go on for ever, and those peculiar people who like mathematics often do. Sometimes they even get quite useful results. It is something like navigating a ship by dead reckoning. If the compass is accurate and you started from where you think you did and you don't drift too much due to unexpected winds and currents and the place you want to get to is where you think it is, and of course if you don't make any mistakes in driving and steering the ship, you get there. It's nicer to be able to see where you're going but in the last resort it's better than doing nothing.

But let us stop at acceleration and try to find out if this idea of a change in the change of distance over a period of time has any practical use.

Distance is something which man has appreciated ever since our ancestors made a flying leap from one tree branch to the next. Speed is something which man has appreciated in a vague kind of way almost as long as he has appreciated distance, but it wasn't until the seventeenth century, when accurate clocks were invented and ships began to sail around the world and cannons began to be used, that speed assumed any precise and accurate meaning. Before that there were really only two kinds of speed, slow and fast. Acceleration came even later in the day, and up till quite recently was more of an abstract scientific idea than something that everyone knew and understood.

One sign of the increasing importance of acceleration in everyday life is the fact that it is acquiring a unit of measurement with a convenient name of its own. Until the first World War the unit of acceleration was the scientific laboratory measurement of so many feet per second per second. Since then, and particularly since the second World War, acceleration is more often measured in "*g*"s, and most people have a vague idea what it means, or at least have heard of the unit.

The main reason for this change is that since the invention of the automobile the ordinary person has been experiencing the results of acceleration and deceleration (which is acceleration in the opposite direction) often in a very painful and personal fashion.

All people who live in a modern industrialized country have some knowledge of automobile accidents. Even if they have not been directly involved they have at least seen the results in photographs in the newspapers. Nobody who sees a photograph of a car wrapped around a lamp post can fail to realize what a tremendous force would have been needed to cause this kind of destruction. What is this force and where did it come from? Most of us realize it was caused by hitting something at high speed but beyond that things are rather vague. Again, if they had to choose between running into either a solid brick wall or a haystack, most people would be quite definite in their choice of a haystack, though apart from a vague idea that the haystack would be "softer" they would have no definite explanation.

The fact of the matter is that the forces that are generated by a sudden stop depend entirely on how sudden the stop is. The people who would sooner run into a haystack than a brick wall have the correct answer, not because the haystack is soft but because it is yielding. Actually, there is quite a difference. For instance, the bumpers on railway trains are not soft as they are made of steel. But because they are spring loaded they yield when they bump into anything and so very efficiently reduce the shock. If we had to run into something, then a steel plate with suitable springs behind it would be better than a haystack.

It is the acceleration and deceleration (or slowing down) which sets up these forces. The greater the acceleration or deceleration the greater these forces will be. It is not the speed but the rate of change of speed that counts.

The question is, how does this tie in with the "g" which we mentioned before? We all know something about gravity. If we lift an object off the ground and then let go, it doesn't stay in mid air but falls to earth again. If we measure the rate of this fall in a vacuum where there is no air to impede the free fall we find that all things from feathers to lead fall at the same rate and that this rate is an acceleration of about 32 feet per second per second. If we change feet per second into miles per hour we will find that 32 feet per second is about 22 miles per hour. So an acceleration of 32 feet per second per second is the same as an acceleration of just under 22 miles per hour per second. Which means the object could accelerate from 0 to 88 miles per hour in just a shade over 4 seconds. Quite a fast take off!

Even if an object is resting on the ground or is prevented in some way from falling, the force which would accelerate it at this speed if it had the chance is still there and we can measure it. It is, in fact, the weight of the object which is in turn the measure of the amount of the pull of the earth on the object. Let us have a look at the automobile in the sketch below. Its weight is one ton. In the sketch the car is hooked

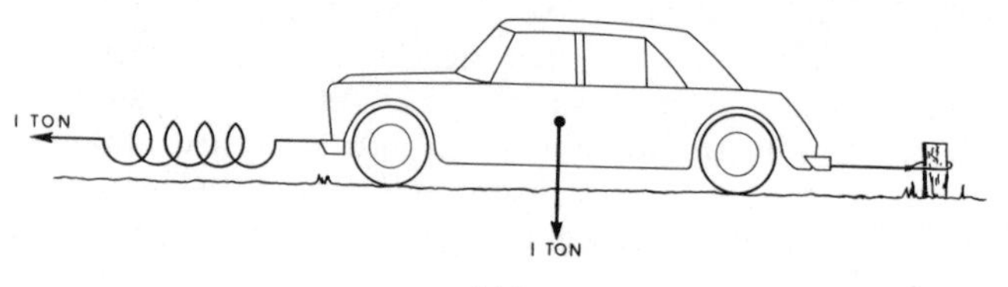

to a post so that it can't move forward and attached to the front of it is a long, strong spring stretched until it is trying to pull the car forward with a pull of one ton.

Now if the rope holding the car to the post were cut, the car would start forward with an acceleration of "g" (that is, an acceleration of 32 feet per second per second) and would keep on accelerating at this rate as long as the one ton pull remained.

Perhaps you are beginning to see the advantages of talking about acceleration in units of "g" instead of in feet per second per second or in miles per hour per second. If it is still not clear, the sketch below may help. The pound weight suspended in the box would not only be pulling toward the center of the earth with a pull of one pound but would also be pulling backwards with a further pull of one pound when the whole box was accelerating. If it were only being accelerated at 16 feet per second per second or $\frac{1}{2}$ g it would only be pulling backward with a pull of half a pound. If it had no acceleration it would of course just hang straight down with no backward pull whatever.

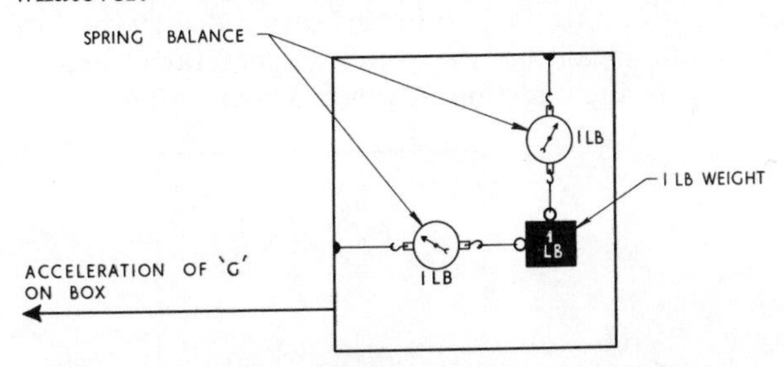

So we see that when there is no acceleration in the horizontal direction we are in a state of horizontal "weightlessness." To get an idea of how the pull of the earth operates we would need something more like the sketch on the next page. Here the pound weight has a horizontal pull of one pound on it, so according to a spring balance anchored to the back of the box it weighs one pound.

Now if we accelerate the *box* forward, the apparent horizontal weight of the pound weight will decrease until, when the box and the weight are all accelerating forward at an acceleration of 32 feet per second per second (that is an acceleration of "g"), the scale registering the horizontal weight drops to zero. Finally, if the box was being accelerated backwards at an acceleration of g and the pound pull was kept on the weight, then the horizontal spring balance would show that the weight was apparently equal to two pounds. We have exactly the same effect when we are going up and down in an elevator. When we are accelerating upward in an elevator our weight increases and when we are accelerating downward our weight decreases. When the elevator is going upward or downward at a steady pace then our weight is normal. It should now be clear why it can be very convenient to use the unit g for acceleration instead of so many feet per second per second. When an object moves with an acceleration of $1g$ it "weighs" its own weight in the opposite direction to which it is travelling; when it is accelerating with an acceleration of $2g$ it weighs twice its own weight; and so on. Deceleration, or slowing down, is acceleration in the opposite direction, so in this case a deceleration of, say, $3g$ would mean that the object weighed three times its own weight in the direction in which it was travelling.

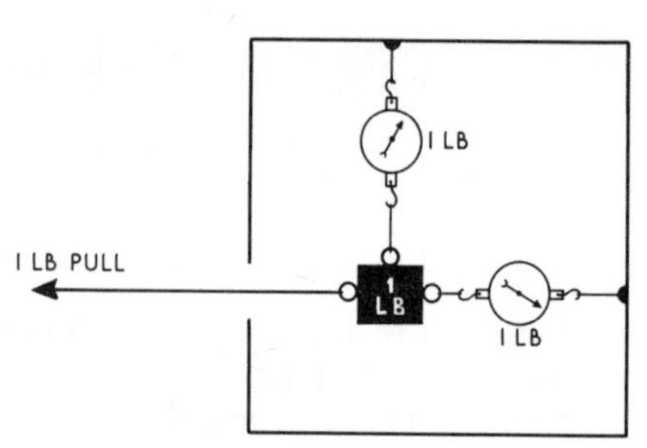

Exactly the same ideas apply to up and down motion except that on the surface of the earth we have a permanent loading of $1g$ downward. So the weightless point comes not when we are standing still but when we are falling, or, to be more accurate, when we are travelling downward with an acceleration of 1*g*. It is interesting to note that if you were travelling downward with say an acceleration of 2*g* you would weigh your own weight *upside down*, that is, in an upward direction. The normal weight is neutralized by half of the 2*g* and the other half sets up a weight in the opposite direction.

The fact that with a downward acceleration greater than 1*g* we get a weight acting *upward* explains among other things why water will stay in the bottom of a pail if we swing it fast enough over our heads even though the pail is upside down. Try it, but don't put too much water in, just in case. If we are accelerating upward, then the weights add together. For instance, if we were accelerating upward at an acceleration of 2*g* we would weigh twice our weight, due to acceleration, plus our normal weight, due to the pull of the earth, which would give us a total of three times our own weight. Everything we held or wore would weigh three times as much and life would be rather uncomfortable. This is the kind of problem that spacemen come up against when their rocket is accelerating away from the earth.

Tests have shown that a man lying on his back on a special couch contoured to the shape of his body could probably withstand an acceleration up to 20*g*. This would mean that a person normally weighing 200 pounds would have a total effective weight of 4,000 pounds. His arm, which might normally weigh about 10 pounds, would weigh 200 pounds.

Now it may be clearer what acceleration and deceleration have to do with car crashes. We pointed out before that if you accelerated from standstill to 80 m.p.h. in four seconds you would be pressed back against the seat with a force equal to your own weight. If you slowed down at this rate you would be pushed forward with the same force. If you want to know what it would be like all you have to do is get your car, stand it on its nose, and get a couple of strong friends to lift you into position in the driver's seat, and then let you drop. The bumps and bruises you get will be much the same as you would have if you slowed from 80 m.p.h. to stop in four seconds.

We can work out that in these four seconds the car would have travelled about 250 feet. In an accident you would be more likely to stop in two or three feet, and the deceleration would be about 100*g*. In this case the effective weight of a 200-pound person would be about eight tons, about eight times the normal weight of the whole car, and every bit of his body would be a hundred times heavier, and pressing on every other part of the body.

At these rates of acceleration living things would squash under their own weight and even an object like a half-inch solid iron bar projecting for a couple of feet would bend double under its own weight. If, however, we have 5 feet to pull up in instead of, say, $2\frac{1}{2}$ then all these forces would immediately be halved.

There is quite a simple formula which tells what the acceleration will be when the speed and the distance travelled from full speed to stop (or from stop to full speed, it is the same acceleration) is known. The formula is simply

$$A = \frac{V^2}{29.75D}$$

where A is the acceleration in g units, V is the speed in m.p.h., and D is the distance travelled in feet.

It is worth noting that the acceleration depends on the square of the speed and therefore double the speed gives four times more acceleration. Thus, other things being equal, you will be squashed nine times flatter if you hit something at 75 m.p.h. than you would be at 25 m.p.h.

In Chapter 4 we showed how nomographs could be made and on the next page is a nomograph showing speed in miles per hour, distance taken to stop, and acceleration in g units. For those who are contemplating having an accident the nomograph should be invaluable. On the center scale is also a scale showing roughly the chance you have of surviving.

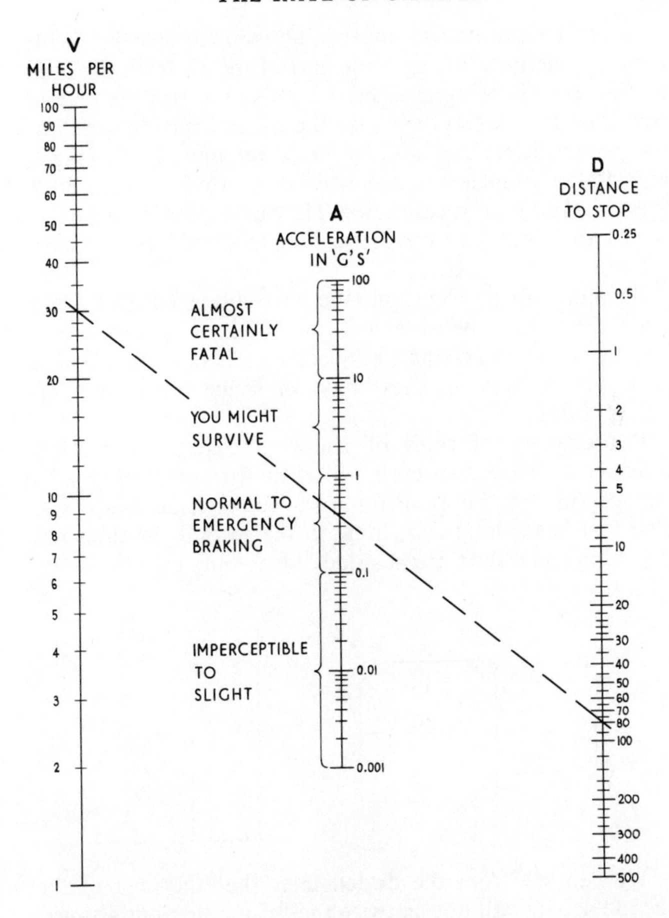

The dotted line represents a straight edge laid to find the acceleration resulting from slowing from 30 miles per hour to stop in 80 feet. The answer on the center scale gives a G of about 0.4, within the normal braking range.

In practice, of course, most injuries in car smashes arise from the victim's hitting some part of the car with his head or through his being crushed and this increases the danger tremendously. Safety belts, on the other hand, by holding the person in position and by stretching and so allowing a few vital extra inches of distance to slow down in, can do a great deal to reduce casualties. The best solution, however, is not to bump into anything. People should try it more often.

So much for acceleration. We have discussed it at some length, not only because it is an excellent example of a change in a rate of change, but because it is something which is useful to know in these days of space travel and fast automobiles.

But changes of rates of change are not confined to movement. They also crop up when we are dealing with changes in size. Suppose for instance we have 3 squares. The first has sides 1 foot long, in the second the sides are 2 feet long, and the third has sides 3 feet long. Like this:

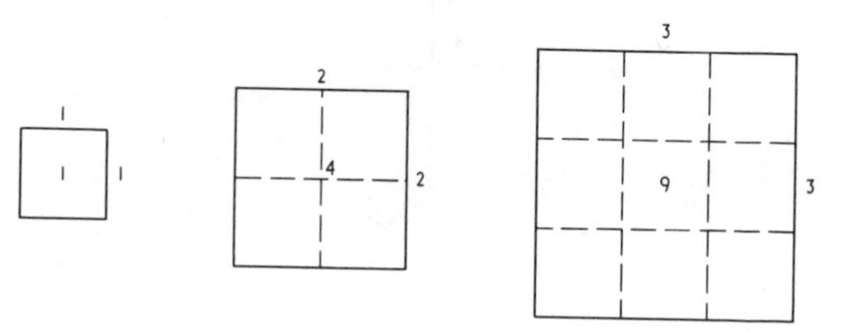

We can see from the dotted lines that the area of the second square will not be twice that of the first but 4 times, and the area of the third square will be 9 times the first, not 3 times.

We can plot a graph which gives a picture of this relationship. If we choose the horizontal scale to represent the length of the sides and the vertical scale to represent the corresponding area, we would have a graph as shown on the next page.

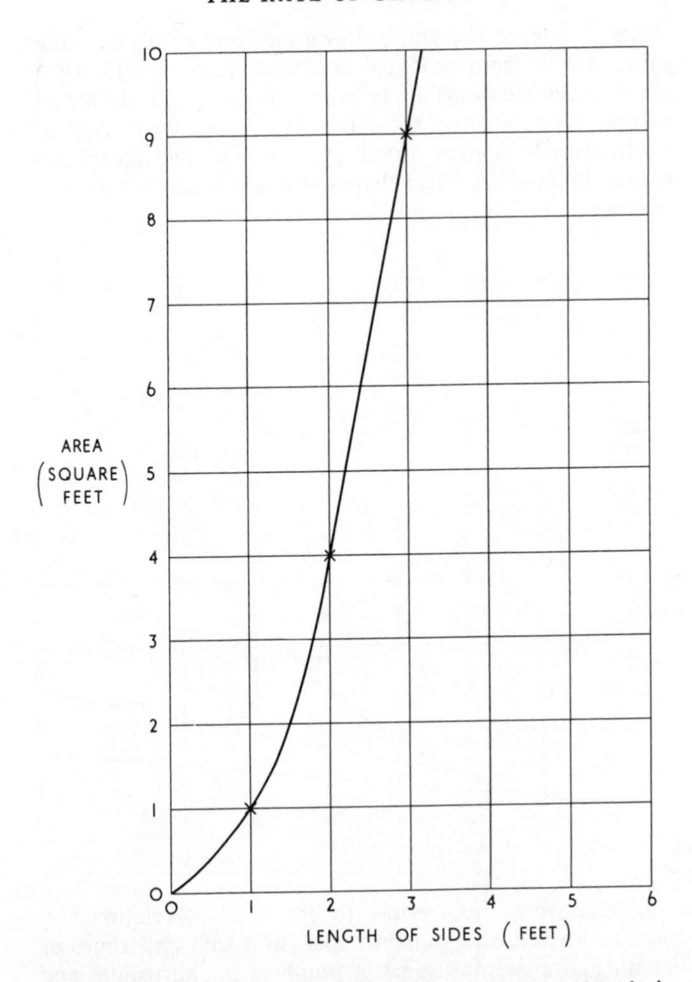

If we examine this graph closely, we will see that it is curved after the pattern of the acceleration graph on page 141. As mentioned before, a very useful property of graphs is that if the quantities plotted have a particular kind of relationship with each other, this will result in a particular shape of line on the graph. If we plot our graph by using values taken from practical experiments, as we did when plotting the relationship between the pull and stretch of a rope, we can often get some idea of the basic type of relationship by noticing the shape of the line. The illustration below shows some basic shapes that are often met with in practice.

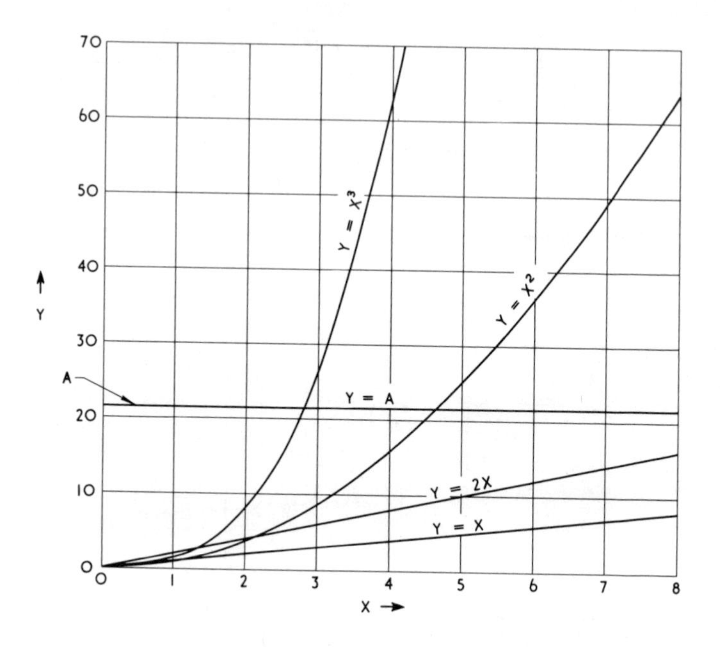

Because these lines apply to any kind of relationship, such as stretch and pull in ropes, or speed and time, or length and area, we use plain numbers for horizontal and vertical scales. And instead of letting the vertical scale represent a particular thing such as distance, stretch, area, pressure, and the like we just call it y as a symbol for any quantity along the vertical scale. In the same way we use x as a symbol for any quantity along the horizontal scale.

On page 139 we showed a graph plotting speed against time when the speed was constant. It was a straight line parallel to the horizontal scale line. In the illustration on page 152, the line marked "y equals a" is the general example of this kind of line. a can represent any fixed number. Since the height up the vertical y scale is independent of the length along the horizontal x scale we find that when x equals 0 y equals a, when x equals 1 y equals a, when x equals 2 y equals a, and so on.

The next general type of line is the one we get when y is directly proportional to x. We saw examples of this in the conversion graph, in most of the other graphs in Chapter 5, and also in the graph showing speed on page 138. The line in this case is a sloping straight line and the general example above is what we get when y equals x, that is, if x equals 1 unit then y equals 1 unit, if x equals 2 then y equals 2, if x equals 3 then y equals 3, and so on.

When the relationship between the two quantities becomes more complicated than this, the graph line ceases to be straight and becomes curved. We can see that this must be so, for the slope of the graph line represents the rate of change of one quantity relative to the other. If this rate of change is changing, the slope of the line is changing, which of course means that it is curved.

One example we have just mentioned is the relationship between the side and the area of a square. The area increases proportionally to the square of the side. In the general graph we show this as y (representing the area) equals x^2 (representing the side squared). The graph of a uniform acceleration, when plotted as a relationship between time and distance, also has this shape. In this case distance is proportional to the time squared.

The next logical step is a graph line representing a change in the change of the rate of change. One instance of this is the relationship between the volume of a cube and the length of one side. If the side is one unit long, then the volume is one cubic unit (that is, $1 \times 1 \times 1$). If the length of the side is 2 units, the volume is not 2×2 but $2 \times 2 \times 2$

which is 8 cubic units. If the length is 3 units, the volume will be $3 \times 3 \times 3$ which is 27 cubic units, and so on.

If we had a graph with the vertical scale representing distance and the horizontal scale representing time, and plotted on it a line representing a uniform change of acceleration, we would get a similar kind of graph, where the distance would be proportional to the cube of the time (that is $time^3$).

The general pattern of this kind of relationship is represented by the line labelled "y equals x^3."

We could go on for ever in this way showing lines representing $y = x^4$, $y = x^5$, and so on. But enough has been said to give the general idea. Those who are interested will find illustrations in standard textbooks which show the kind of shape one can expect from various relationships and some uses which have been found for them.

The illustrations in most textbooks also show something which we have not mentioned, namely, what happens when the units along the horizontal scale (that is the x scale or x axis) become negative. In the case y equals x, y becomes negative, but in the case of y equals x^2 y remains positive. In the case of y equals x^3 the y units become negative. This is all very well as long as x and y represent the relationship of pure numbers but it doesn't mean that a cube with sides in one particular direction has a negative area. These "general" relationships are not so general that we can apply them without a little forethought, or else we will get into the "square sheep" department; which again shows the need to treat mathematics with a lot less respect and a lot more caution.

Finally, there are two ratios concerned with change which are so important and fundamental that they will have a special mention in a later section. In addition, some of the relationships shown in this chapter will crop up in later parts of the book. Have a good look at them and get to know what they look like so that you will recognize them again. Whether you are a policeman, a bouncer in a night club, or a mathematician, the success you achieve in your profession will often depend on your ability to recognize old faces in new and unexpected situations.

NOTE: Throughout this section the familiar word "speed" has been used instead of the correct word, which is "velocity." Velocity could be described as speed in a straight line. The forces concerned and acceleration itself come from changes in velocity, and arise even if the speed is constant, as you can easily discover by going around a sharp bend at a constant speed of a hundred miles an hour.

8. THE RIGHT ANGLE ON TRIGONOMETRY

My grandpaw was the best hand at triggernometry I ever seen. He could shoot the center out of the Ace of Spades at a hundred paces.

Memoirs of a Hillbilly

The average person does at least realize that trigonometry has something to do with triangles, or "trigons" as they are sometimes called. (If you want to make things difficult and confusing, there is nothing better than giving the same thing two or three different names!) Beyond this knowledge, however, it is a subject about which he tends to be rather wary. Anyone who still has memories of ploughing through Euclid's eternal triangles in his schooldays will sympathize with this point of view. Fortunately for our purposes, there is no need to resurrect any of these bygone tribulations. The amount of trigonometry needed to give a basic understanding is no more than can readily be absorbed by any reader with an average amount of common sense.

The trigonometry we are going to talk about here is based on right-angled triangles. We have come across these already and know that they have the usual three sides and one angle which is a right angle of 90 degrees. Like this:

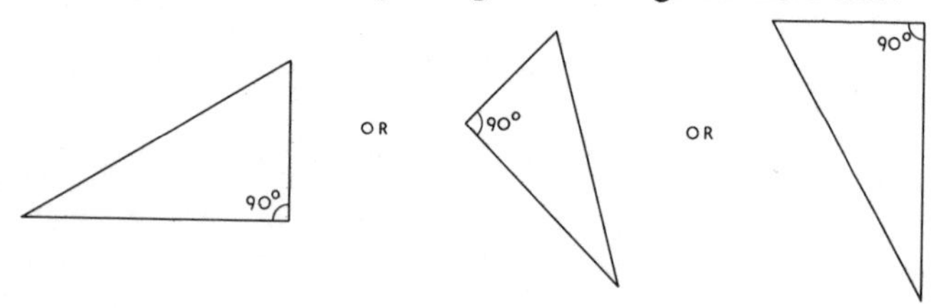

It doesn't matter which way we draw our triangle as long as one of the angles is a right angle.

Suppose now that we draw three triangles, each of a different size, but all having the same angles. Like this:

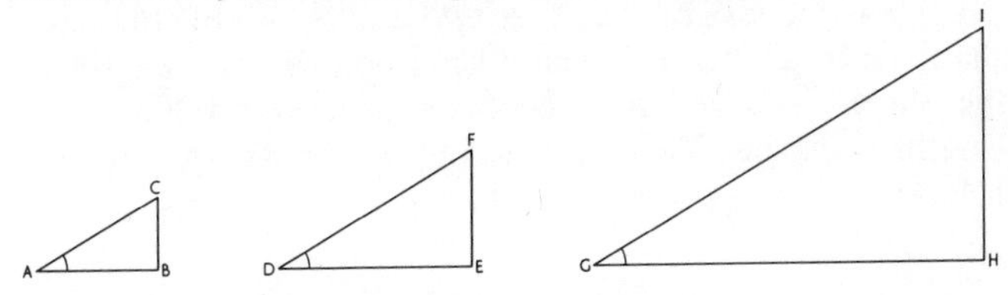

If we put these triangles one over the other they would fit together because the angles at *A*, *D*, and *G* are all equal. Like this:

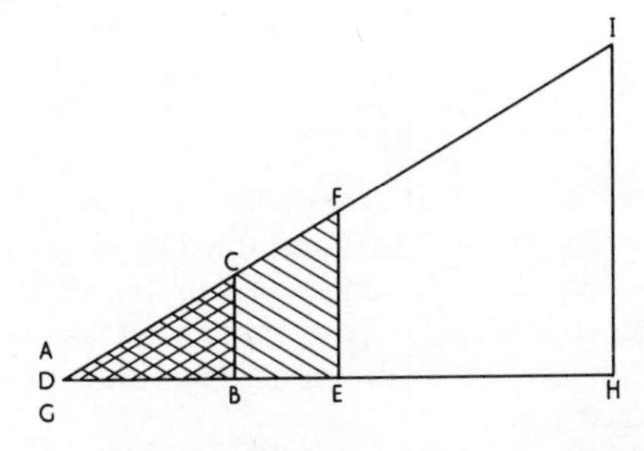

If we study these two illustrations for a little it will be obvious that although these triangles are not the same size, they have the same basic shape. In other words, if the side *AC* of the smallest triangle is a quarter the length of side *GI* of the largest one, then *AB* will be a quarter the length of *GH* and *BC* will be a quarter of side *HI*. If *DE* were twice *AB*, then *EF* would be twice *BC*, and so on.

Another way of putting it would be to say that the ratio between *BC* and *AC* of the smallest triangle is the same as the ratio between *EF* and *DF* of the center triangle and sides *HI* and *GI* of the largest one. The smallest triangle is in fact something like a scale model of the larger ones—all its proportions are the same though its size is different. Those who want to go into a "proper" proof of these things can find them in any standard textbook.

The practical importance of all this lies in the fact that the only thing that determines the shape of a right-angled triangle is the size of the other two angles. And if the basic shape is fixed, then the relationship between the lengths of the three sides will also be fixed, whatever their actual lengths may be. Take, for example, the triangle shown below:

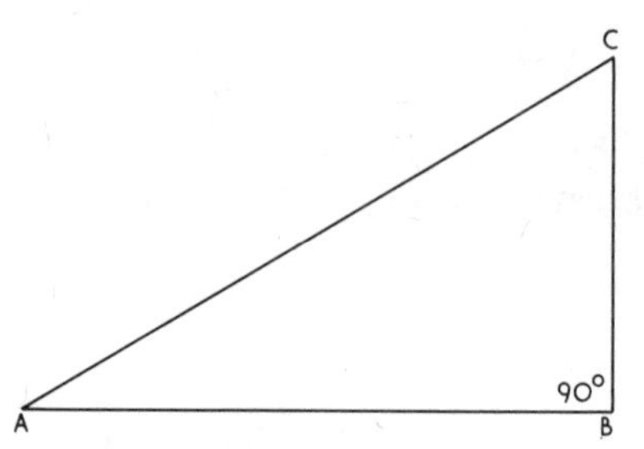

As long as the angle at A remains the same, the ratio of side BC to side AC will be constant. Or, to put the same thing in another way, the length of side AC divided by the length of side BC will always give the same answer as long as the angle at A remains the same, regardless of the actual size of the triangle.

It is this fact which makes it possible for us to do all kinds of survey work, and navigational and astronomical calculations quickly, easily, and accurately. The illustration below will give a simple example of how this fact can be used.

Suppose we wish to measure the height of a tree. If we wait until the sun is shining, and measure the length of the shadow, then put a stick upright in the ground and measure

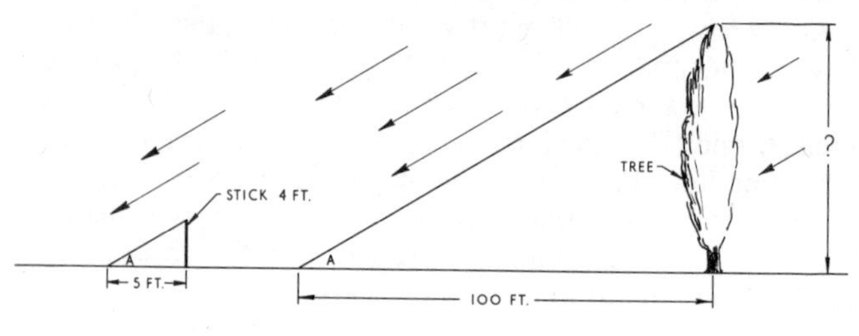

its height and also the length of its shadow, we have all the information we need. The ratio of the stick's shadow to the height of the stick is the same as the ratio of the tree's shadow to the height of the tree. In the illustration we see that the length of the stick's shadow is 5 feet, while the height of the stick is 4 feet. So the height of the stick is 4/5 times the length of its shadow, or the length of the shadow is 5/4 (or $1\frac{1}{4}$) times the height of the stick (which, as we saw when discussing ratios, is another way of saying exactly the same thing).

Because the angle marked "*a*" in the illustration is the same for both the stick and the tree it follows that the height of the tree is also 4/5 times the length of its shadow. Since the tree's shadow is 100 feet, its height must be $100 \times 4/5$ or eighty feet.

Now one of the undeniable advantages of mathematics is that, within limits, it makes it possible for us to apply information learnt from one situation to a different situation, without having to start from scratch every time. And once we have learnt, by our stick and shadow, or any other way, that a triangle which has a certain angle has sides with a certain ratio, there is no need to learn it afresh every time. We can make up a table showing various angles and the corresponding relationships of the sides. In such a table we would record that any triangle with an angle the same as our angle marked "*a*" would have a height to length ratio of 4 to 5.

So if we go out to measure the tree again, we don't have to worry about the stick and its shadow, or even the shadow of the tree. We would just measure back a convenient distance from the tree (say 100 feet just to correspond with the first illustration) and then measure the angle between the horizontal and the top of the tree which we would find to be equal to "*a*". See sketch on the next page.

We would then look up our table and say, "Ah yes, for an angle the size of "*a*" the height to length ratio is 4 to 5, so the height of the tree must be 4/5 times 100 feet, which is eighty feet."

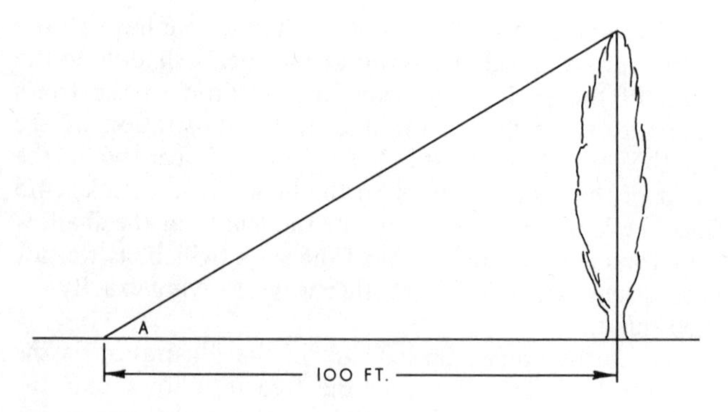

No fuss, no bother, no sticks or shadows. Only a table of angles and their ratios.

One snag is that we have to measure an angle, but actually this is a lot easier than it sounds—even easier than measuring a length.

There are gadgets such as theodolites which are especially made for this kind of job. The principle on which they work is fairly simple. The bare essentials are shown in the sketch below.

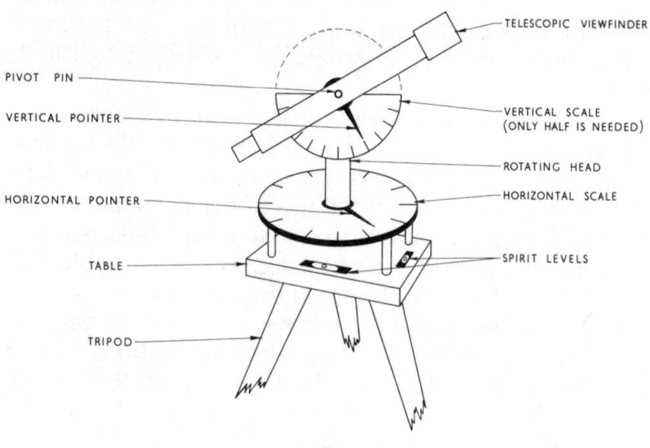

The base consists of a small table mounted on a tripod and having two spirit levels on it so that it can be adjusted until it is absolutely horizontal. This enables us to have a horizontal reference point to start from, even if the ground is uneven or hilly.

On this table is a horizontal circular scale similar to a school protractor. This scale is marked in degrees around the full circle starting at 0 and finishing at 360 degrees. Sticking up through the middle of this is a vertical post on which the upper part of the head of the gadget can rotate. This upper part has another circular scale much the same as the first one, except that it is vertical, and sticking horizontally through the center of this is a pivot pin on which the telescopic viewfinder, with the vertical pointer fixed to it, can swing.

To find our angle "*a*" with a theodolite, all we would have to do is to level it up, swing the telescopic viewfinder until the top of the tree is in the center of the viewfinder, and then read off the angle on the vertical scale.

In practice one has to allow for the slope of the ground when measuring the horizontal distance, and also to allow for the fact that the pivot pin of the theodolite, which is the point from where the angle is measured, is not on ground level. In really fancy survey work people even take into account the possibility of the light rays being bent by pockets of hot air (the shimmering watery appearance of a hot roadway is an example of light rays being bent by hot air) and a host of other things. But unless we aim to become surveyors we need not worry about these details.

The method of getting distances by measuring angles is just as useful for getting horizontal distances as it is for vertical ones. Let us look at the sketch on the next page.

Suppose we wanted to find the distance between two points divided by a steep gorge which is too wide to be spanned with a tape. All we have to do is name the points. *A* and *C* and choose another, *B*, on the side which we stand. The distance between *A* and *B* is such that we can measure it accurately.

We then set up the theodolite at point *A* and get the angle between line *AB* and line *AC*. Then we move to *B* and get the angle between line *AB* and line *BC*. Then we pack up and go home, having got all the information we require, without even having had to get across to the other side of the gorge.

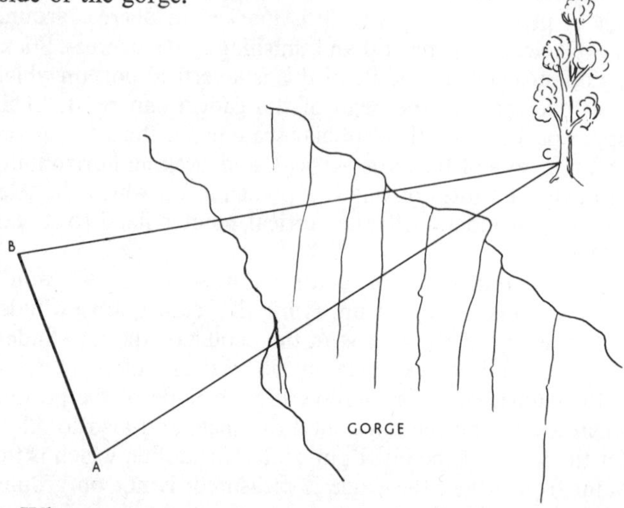

When we try to work out the length of sides *BC* and *AC*, however, we run up against a snag because we wake up to the fact that the triangle is not a right-angled triangle, and therefore our convenient ratios don't directly apply. We can, of course, by carefully choosing pòints *A* and *B* and *C*, make sure we do get a right-angled triangle, but it is not necessary. We can still get an answer from the information we already have, and there are two ways in which it can be done. The first way is simply to draw it to scale. If the base line was 100 feet, we could draw a line, say, 10 inches long and let it represent the base line *AB* at a scale of 10 inches being equivalent to 100 feet (which is a scale of 1 inch equals 10 feet). Then since we know the size of the angles at points *A* and *B* we can draw lines from these points at the correct angles. The place where these lines intersect each other will be our point C at a scale of 1 inch equals 10 feet. Like this:

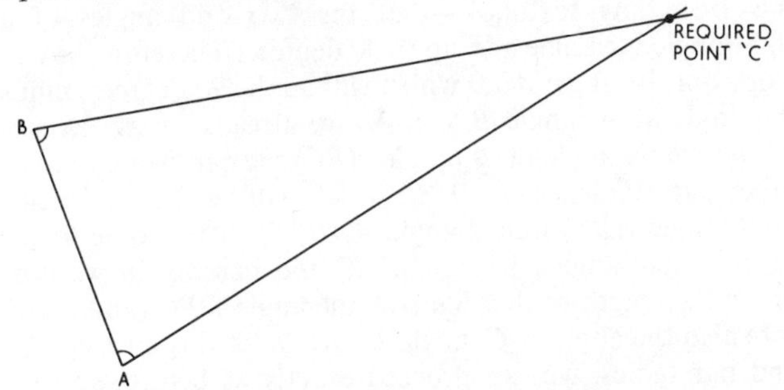

The main drawback about this method is that its accuracy depends on the accuracy with which we draw the base line and set out the angles. And sometimes a small error in setting out the angles can give a large error in the place where the lines intersect, and so in the position of point C. On the whole it is better to work it out using the information we have about right-angled triangles. The fact that we haven't a right-angled triangle doesn't matter in the least. We simply go ahead and make one, or rather a pair of them, like this:

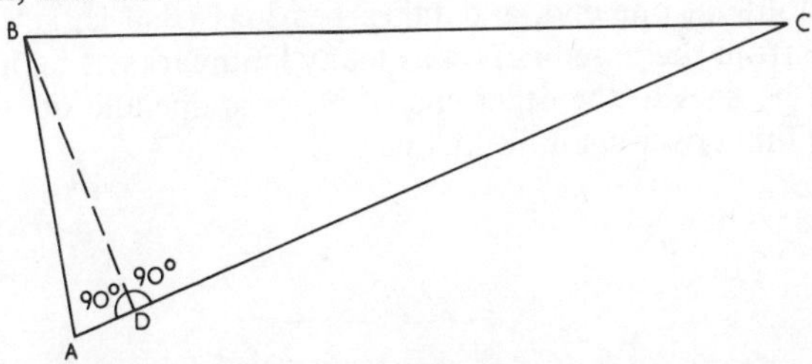

We simply imagine (we don't have to draw it accurately) that there is a line from B which meets the line from A to C at right angles at point D.

Now we do have a right-angled triangle ABD especially made to order, and we know the size of the angle at A (that is, angle BAD) and also the length of the line AB. So from our tables we can find the length from *A* to *D* and also from *B* to *D*. We also know—you can get the proof of this from any textbook—that the two odd angles of a right-angled triangle add up to 90 degrees. Therefore we can work out the angle *ABD* which will be the 90 degrees minus the angle at *A* (angle *BAD*). As we already know the size of the whole angle at *B* (angle *ABC*) we can work out the other part angle, angle *DBC*, which will be the total angle at *B* (angle *ABC*) minus angle *ABD*. Then we come to the second right-angled triangle *DBC* and here again we now know the length of the side *BD*, the angle *DBC* (and therefore also the angle at *C*, angle *BCD*). With this information and our tables, we can proceed exactly as before, and get the length of side *BC* and also of side *DC*. The total length of side *AC* of the original triangle is then the sum of length *A* to *D* and *D* to *C* both of which we now know. And thus we have found out the position of point *C* in relation to points *A* and *B* without having had to measure it directly.

It is easy to see how useful this method can be in making a map of a district, particularly if there are rugged mountains and lots of gullies and rivers. The surveyor simply selects a suitable flat area and lays out a base line which he can measure without any trouble. Then he sets up his theodolite at one end, and takes bearings (that is, gets the angle from the base line) on as many landmarks as he likes. Then he goes to the other end of his base line and does the same thing over again. Like this:

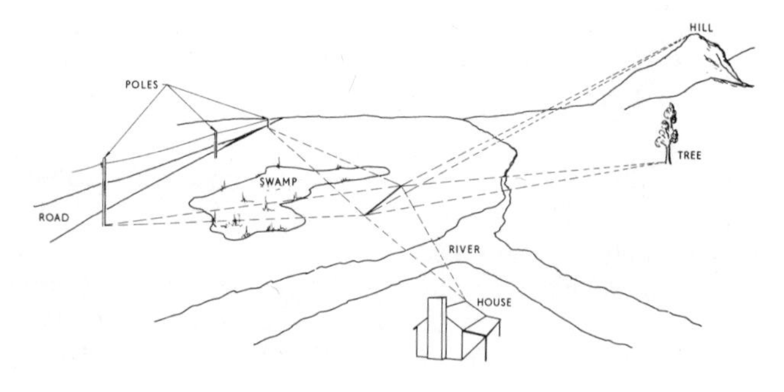

Again, in real life things are a little more complicated, because he has to worry about triangles on edge for getting heights and triangles on the flat for getting horizontal distances at one and the same time. But again, we don't have to go into all these details in order to understand the basic principles involved.

It has already been pointed out that we can make a table showing what relationship the lengths of the sides have to each other for any particular angle. In a triangle with an angle of 30 degrees the ratio of the sides is like this:

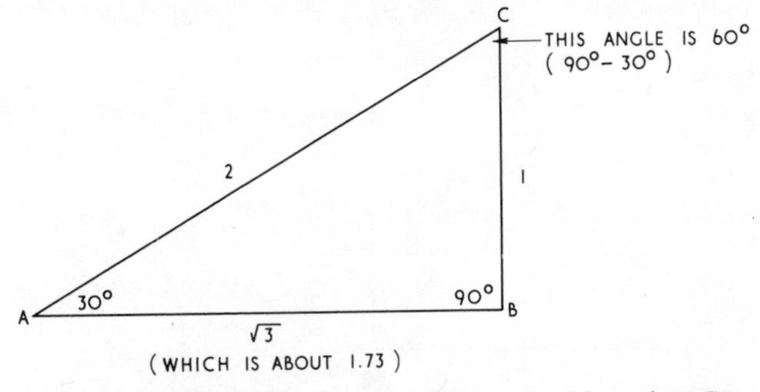

Here the relationship between the vertical length *BC* and the horizontal length *AB* is as 1 is to 1.73.

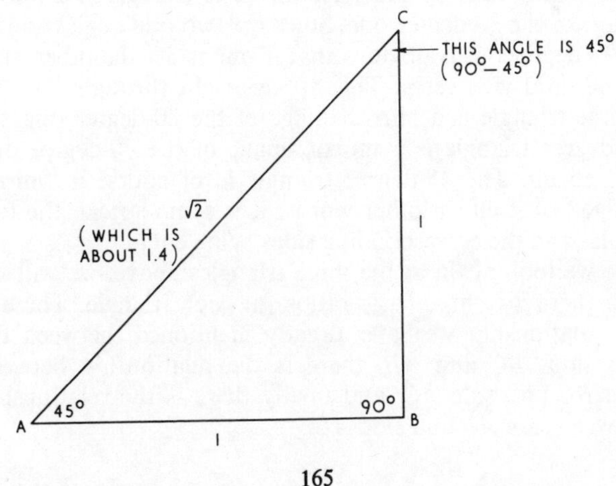

In a triangle where the angle at *A* is 45 degrees the relationship between *BC* and *AB* is one to one.

And if we have a triangle with an angle of 60 degrees we have a relationship like this:

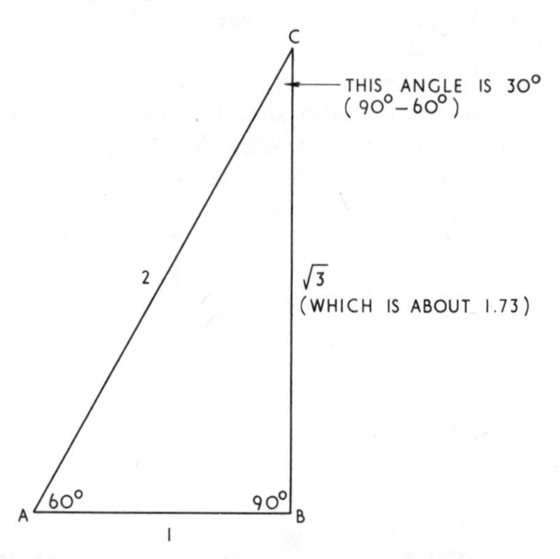

Here the ratio between *BC* and *AB* is as 1.73 is to 1. You will notice, by the way, that this triangle is a mirror image of the 30-degree one. Since the two odd angles add up to 90 degrees, it is obvious that if one is 30, the other must be 60, and vice versa. This applies right through. The 10-degree triangle is a mirror image of the 80-degree one, the 20-degree triangle is a mirror image of the 70-degree one, and so on. The 45-degree triangle is of course a "mirror image" of itself, in other words, it is symmetrical, the two angles and the corresponding sides being equal.

If we look again at the three triangles above, we will see that there are three basic ratios for each triangle. There is the relationship we have already mentioned, between the two sides *BC* and *AB*, there is the relationship between side *BC* and side *AC*, and finally there is the relationship between side *AB* and side *AC*.

The use of the phrase "side *AB*," "side *BC*," and so on is a rather roundabout way of describing them, and in addition it means we have to be looking up all the time to see which side is which. A much better way of describing them is to talk about the side opposite the angle in question and the side adjacent to it. The third side, the longest one, has a name anyway. It is called the "hypotenuse" and the sensible thing is to call it by this name. Our triangles now look like this:

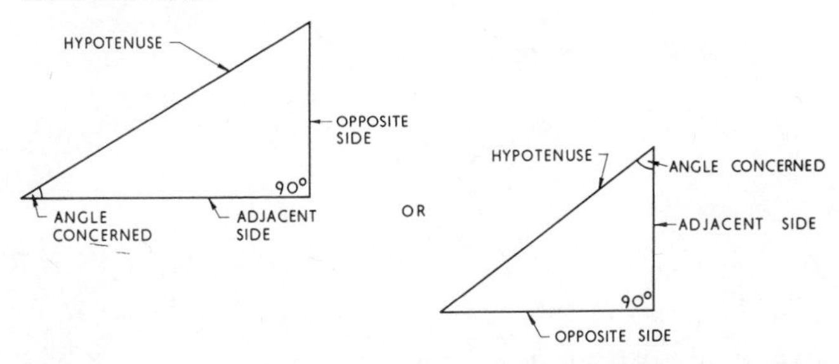

We don't have to worry what way up we draw the triangle or even if we draw it at all. The opposite side is always the one opposite the angle concerned, the hypotenuse is the longest side and is always opposite the right angle, and the remaining side is the adjacent one. So now we can describe the three relationships between the sides in a simpler and clearer way. We have the relationship between the opposite side and the adjacent side, between the opposite side and the hypotenuse, and between the adjacent side and the hypotenuse.

We have already pointed out, when talking about ratios, that if a ratio achieves a high popularity rating through frequent use it often gets a name of its own. We don't say a car is travelling a certain amount of distance in relation to a certain amount of time, we say it is travelling at a certain speed. In just the same way, instead of talking about the relationship between the opposite side with respect to the angle concerned and the adjacent side with respect to the angle concerned, we simply say the "tangent" of the angle concerned. In other words, the ratio of the opposite side divided by the adjacent side is called the tangent.

Mathematics has, like many other languages, an absolute gift for giving confusing and misleading names to things, but in this case the name has a certain amount of sense to it. The ordinary definition of a tangent is that it is a straight line which touches a curved line but does not cross it. Like this:

If we think for a little we will realize that where the curved line and the straight line touch they must be going in the same direction. If they were not they would cross instead of only touching. We can use this idea when we want to measure the steepness at some point on a slope which is continually changing, a situation such as we would get near the top or the bottom of a hill where the slope was flattening out or perhaps getting steeper. Like this:

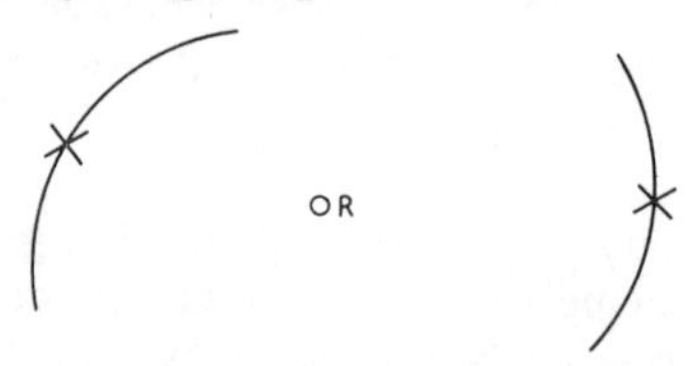

The cross in each case shows the place where we want to measure the slope. We already know that a tangent line drawn through these places will have the same slope as the curve, so the obvious thing to do is to draw a tangent and measure its slope.

THE RIGHT ANGLE ON TRIGONOMETRY

As most of us know, the steepness of a slope is measured by the ratio between the amount travelled vertically and the amount travelled along, or in other words we divide the vertical distance by the horizontal distance. A slope of one in thirty (or one thirtieth) means that we go one foot up for every thirty feet we move in a horizontal direction. A slope of one in one (or more simply a slope of one) would mean that for every foot along we would move a foot upwards also. Thus a slope of one would have an angle of 45 degrees.

The sketch below shows our two curves with tangents drawn at the points where we want to measure the slopes. They now look like this:

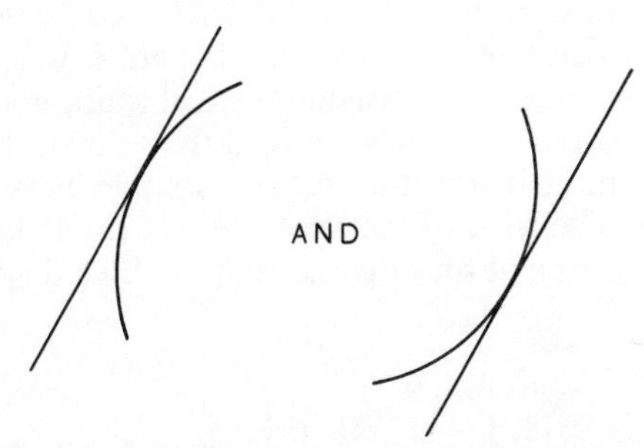

As pointed out above we measure the slopes of these lines by dividing the distance up by the distance along. To get these distances we draw horizontal and vertical reference lines, like this:

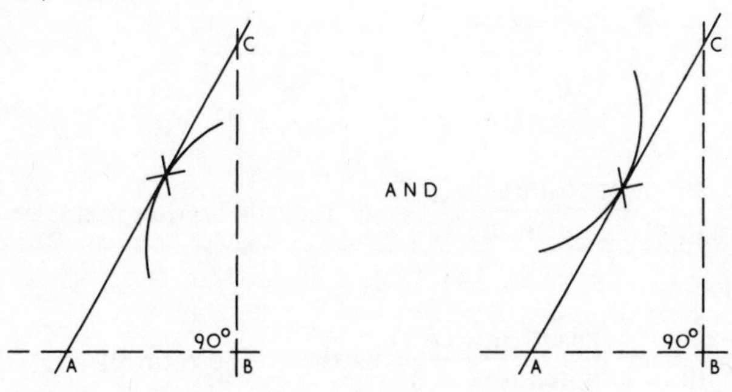

The addition of these reference lines gives us our familiar friend the right-angled triangle with the tangent line as the hypotenuse. And the ratio BC divided by AB is not only the measure of the slope of the tangent line but it is also the ratio which is called in trigonometry the tangent of the angle at A. The angle at A is the angle the tangent line makes with the horizontal reference line, and so we can see there is some genuine relationship between a tangent line and the trigonometrical ratio which is also called a tangent.

Whether it is a good idea to use the same word for two different things is another matter. One is likely to get into the same kind of muddle as the Martian who, finding his girl friend liked to hear her hat described as "ducky," tried to go one better and said that he thought it was fowl.

Like the tangent (the trigonometrical ratio, not the line), the other two relationships between the sides of the triangle have also graduated to the stage where they have names of their own. The ratio of the opposite side to the hypotenuse is called the sine of the angle concerned. Like this:

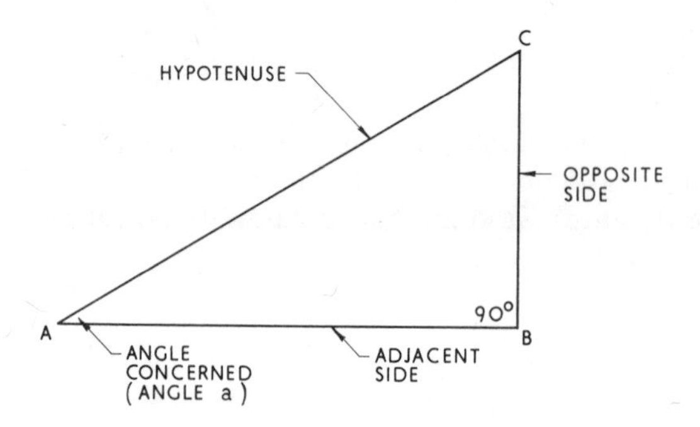

$$\frac{\text{Length of opposite side } (BC)}{\text{length of hypotenuse } (AC)}$$ is the sine ratio of angle a,

also:

$$\frac{\text{length of adjacent side } (BA)}{\text{length of hypotenuse } (AC)}$$ is the cosine ratio of angle a,

and, as we already know:

$\frac{\text{length of opposite side}(BC)}{\text{length of adjacent side }(AB)}$ is the tangent ratio of angle a.

It does not seem clear why the ratio of the opposite side to the hypotenuse was called the sine of the angle. There is a story that it was invented in order to cook up the joke about a surveyor being an unfortunate person who goes around loaded with (measuring) chains, always looking for a sine. This, however, does not seem entirely probable, even for mathematics. A more probable explanation is that it was called after the Latin for bow string since the arc of a circle is involved. It would have a distant relationship to the word, sinew. The third ratio, the cosine, was probably given its name because it co-exists with the sine. . . . That isn't altogether a joke because if we look back at pages 165 and 166 we will see that they are actually very closely linked. In the first triangle the sine of 30 degrees was 1/2 (that is 0.5) while the cosine of 30 degrees was 1.73/2 (which works out to about 0.866). In the next triangle, where both angles are 45 degrees, the sine and cosine are both the same value, 1/1.4 (which comes to about 0.707). Finally, when we get to the 60-degree triangle, the values of the sine and cosine have been reversed. The sine is now 0.866 while the cosine is now 0.5.

The reason why the sine and cosine are so closely linked with each other will be clear from the sketch below.

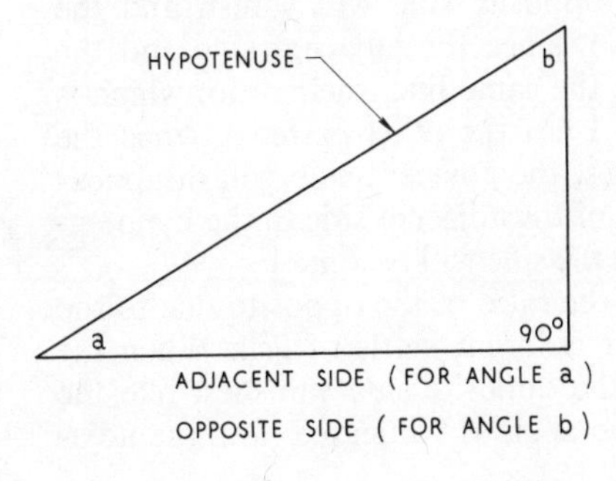

The side which is the opposite one for angle *a* becomes the adjacent side for angle *b* and vice versa. And since these two angles always add up to 90 degrees it means that the sine of any angle has the same value as the cosine of 90 degrees minus that angle. So, for example, the sine of 10 degrees is the same as the cosine of 80 degrees. The cosine of 40 degrees is the same as the sine of 50 degrees and so on.

Finally, it would be useful to know what actual values we can expect these ratios to have for different size triangles. It is not very difficult to do this because we only have to worry about angles between 0 degrees and 90 degrees. A glance at the sketch below will show us why.

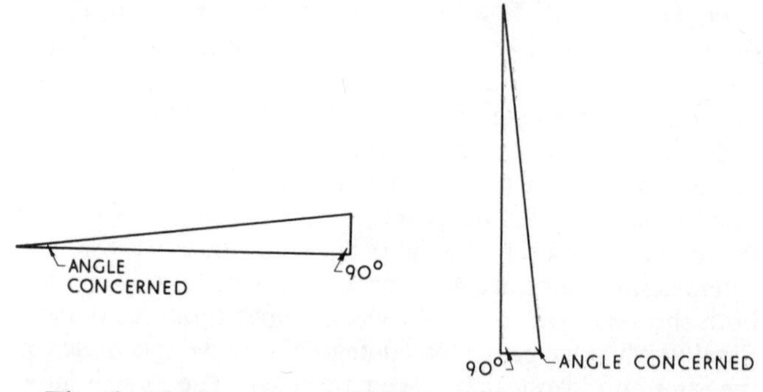

The sketch shows two triangles. In the first one the angle concerned is approaching 0 degrees and in the other one it is approaching 90 degrees. It is clear that when the angle becomes 0 degrees the opposite side will vanish and the sine will become 0. Also, since the adjacent side and the hypotenuse will become the same line, their ratio, which is the cosine, will become 1. At the other extreme, when the angle becomes 90 degrees, the adjacent side will disappear and the cosine, the ratio of the adjacent side to the hypotenuse, will become 0, while the sine will become 1.

The tangent, which is the ratio of the opposite side to the adjacent one, is a rather different kettle of fish. When the angle is 0 degrees and the opposite side vanishes, like the sine it is 0. But when the angle is 90 degrees and the adjacent side vanishes, the value of the tangent becomes 1/0 and what that means—apart from infinity—is anyone's guess. As may be expected, the value of the tangent gets very large over the last few degrees. At 80 degrees it is about $5\frac{1}{2}$, at 89 degrees it is about 57, and at 89.9 degrees it is up to 573 and still going strong. When we remember that the tangent ratio is the measure of the slope (or steepness) of the hypotenuse, these figures make sense; for at 0 degrees, when the hypotenuse is lying along the horizontal base line, its slope is nothing. When the angle is 90 degrees the hypotenuse is standing up vertically and its slope is infinitely great.

That is the basic story of trigonometry without any trimmings, and if we were confining ourselves to well organized and properly arranged triangles on a sheep-and-cow level of mathematics, we could wrap it up and be done with it. But just as the sheep-and-cow numbers and methods and symbols have been put to completely new uses in the mathematics of measurement, so the sines and cosines and tangents have been dragged out of their nice comfortable triangles and put to uses which the people who first invented trigonometry never dreamt of. If you don't have a liking for mathematics, then the less you know about some of these uses the better; but there is one development which is of such wide practical importance in the present-day world that we could hardly ignore it even if we wished.

You will remember how negative numbers, which were meaningless in sheep and cow mathematics, came to have a fundamental significance in the mathematics of measurement. There the minus sign became a label which indicated direction, a term which had no meaning in sheep and cow mathematics, because plus one sheep is plus one sheep, whether it is lying on its back or standing on its head.

In exactly the same way, we find that direction has no meaning in the ordinary geometry of Euclid and the trigonometry which is based on it. According to ordinary geometry a triangle like this:

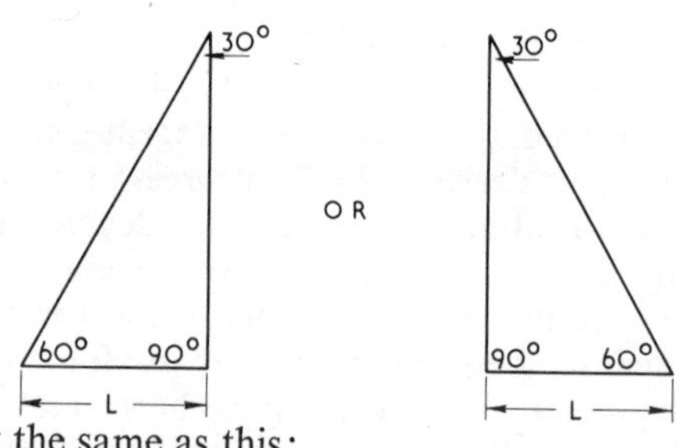

is exactly the same as this:

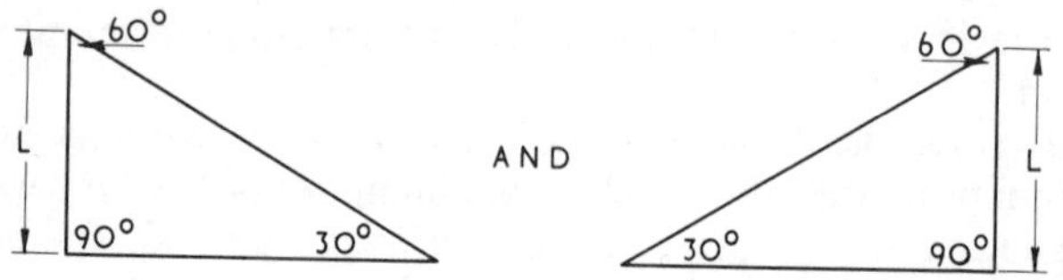

provided that the angles and the sides are the same. In trigonometry based on these triangles the same thing applies. The sine of 30 degrees is $\frac{1}{2}$ whatever way the triangle happens to be sitting.

This kind of thing was quite satisfactory for building pyramids and surveying to a limited degree and for working out the height of trees and cliffs and the lengths of diagonal pieces of timber in buildings. Suppose however we bring these triangles into line with the rest of the mathematics of measurement, where things have direction as well as size. In maps and graphs we have always assumed that a direction up and to the right is positive. For instance, when we have talked about a slope of $\frac{1}{2}$ we have shown it like this:

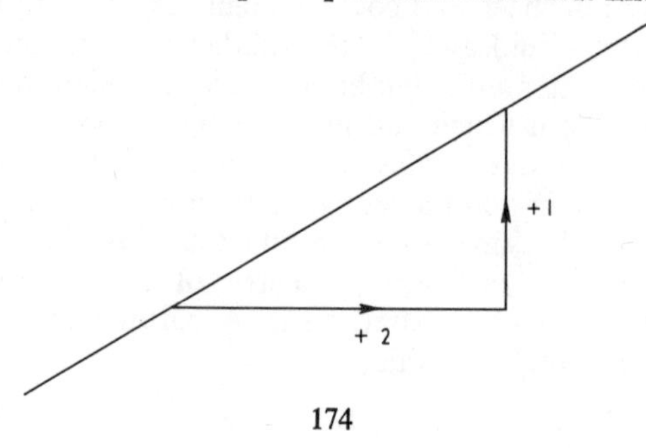

As was mentioned earlier, when one sees any number or symbol, like 1 or 3 or "x" or "a," we really mean plus 1, plus 3, and so on. It seems an awful waste of time always writing "plus" in front of everything when it isn't immediately useful, but if we don't we are very likely to forget and begin to imagine we are dealing with some kind of disembodied phantom figure which is neither plus nor minus. To get back to business, however, the fact is that if we say that a direction to the right and a direction upwards are positive (plus) we must say that the opposite directions are negative, like this:

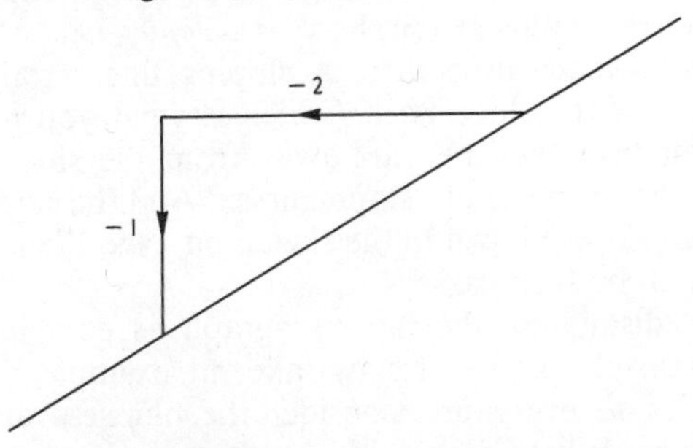

Here we have a slope, not of one in two or 1/2 but a slope of minus one in minus two or −1/−2. The most interesting thing about it is that it is the same slope. This ties up with the mathematical side of the business because we find that minus one divided by minus two gives plus one half as an answer, just as does plus one divided by plus two.

This is where life can get a little complicated. "Surely," someone will ask, "if a line in one direction is positive, and a line in the opposite direction is negative, shouldn't a slope in the opposite direction be negative? Like this?"

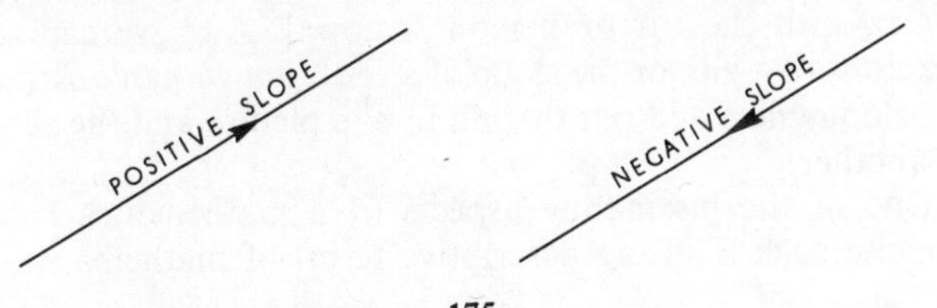

Although this interpretation looks plausible it happens to be quite wrong!

If it was correct, then our minus one divided by minus two would have to give an answer of minus one half to be consistent, but it most certainly doesn't. And this is one interesting case where mathematics is right and our first impressions can be wrong.

What our illustration shows is a length and a direction, not a slope. A slope, as was explained earlier, is a ratio, one of those awkward things that are not things at all. It is the measure of the amount we go up to the amount we go along. What the illustration shows is a *sloping line*, which is a different business altogether. A sloping line certainly has direction but the slope itself (which is what you have left when you take the line part away from the sloping line) has only the property of "slopingness." And the slopingness of the two slopes shown in the sketch on page 175 is exactly the same in both cases.

Before dismissing the last paragraph as completely incomprehensible jargon, let us take an example that will be familiar to everyone. Consider the blueness of a blue automobile. What is this blueness? It isn't an inherent part of the automobile because all automobiles aren't blue. It isn't an inherent property of the paint because paint doesn't have to be blue. The blueness could be described as what is left after we take the automobile, the paint, and everything else away, like the Cheshire cat's grin which remained after the cat itself had vanished. Here again we have ideas setting themselves up and pretending to be things. The fact is that we can see a blue car or a blue wall or a blue something but we can't see a blue. We can see a grin on a person or even on a Cheshire cat but we can't see a grin with the cat or person removed. And we can see the slope of a hill, or the slope of a roof, but we can't extract the slopingness and put the hill in one picture and the slope in another.

One of the fascinating aspects (if a mathematics hater can use such a strong descriptive term) of mathematics is that in it one can take ideas, treat them as though they were actual things, manipulate them, and often get results which years later provide the key to some important discovery. (On the other hand, one can also get, after a lot of hard work, a profound and comprehensive analysis of the properties of cubic cows or square sheep.) Thus, while we can't separate in a picture the slopingness of a slope from the slope itself, we can do so in mathematics.

Getting back to the measurement of this slopingness, we found that a slopingness of one in two and also of minus one in minus two was the same thing, namely $\frac{1}{2}$, which is plus.

Is there such a thing as a slopingness of $-\frac{1}{2}$ and if so what is it? From ordinary arithmetic we know there are two divisions that will give us a negative or minus answer, namely a plus quantity divided by a minus one, or a minus quantity divided by a plus one. So again we have two combinations, like this:

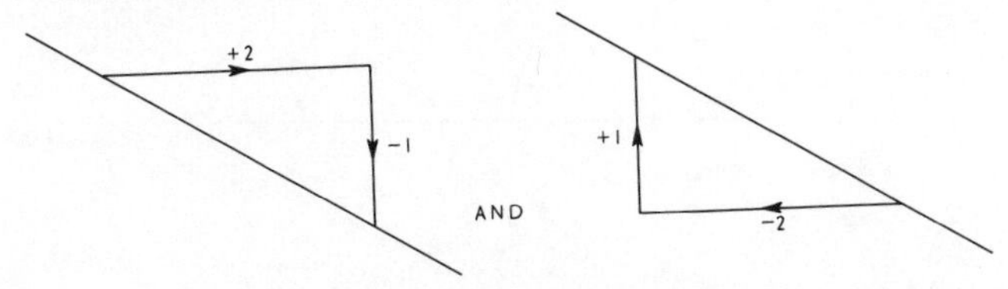

We see therefore that there is a negative slope which has the opposite kind of slopingness (*not* the opposite direction, for slopes don't have direction) to that of the positive slope.

Having added the properties of direction to the "up" and the "along" sides of our triangle and having found that this has resulted in the third side acquiring two possible kinds of slopingness, we can turn our attention to putting direction into the angle. Suppose we have two angles:

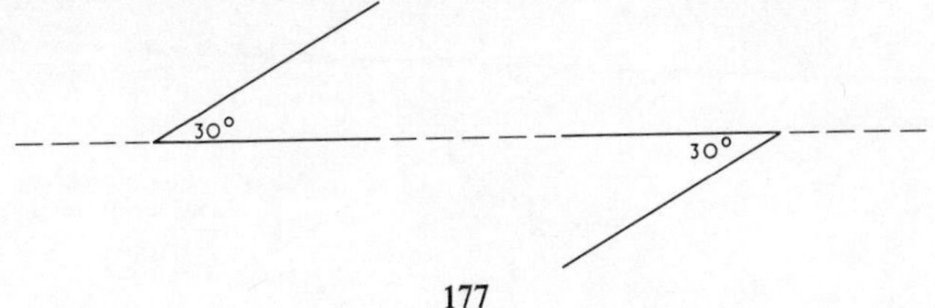

In ordinary trigonometry we would say they were identical angles, just as we regard a line as having only length but not direction. To introduce direction we think of an angle as being created by a line swinging around on a pivot point, the pivot point being the apex of the angle. Like this:

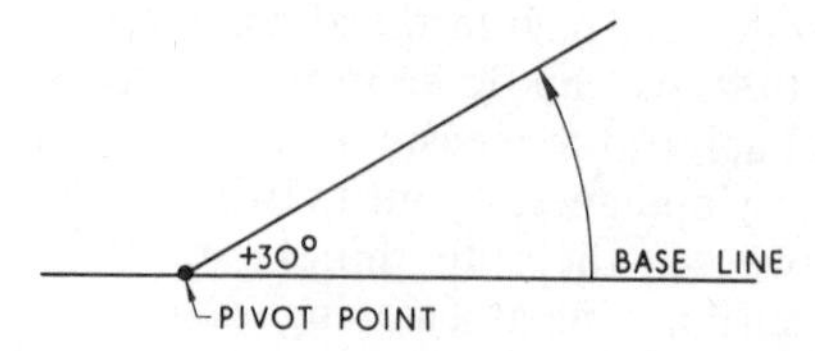

Now if we consider the angle shown above as being plus 30 degrees we must consider an angle swept out by rotation in the opposite direction as having a value of minus 30 degrees.

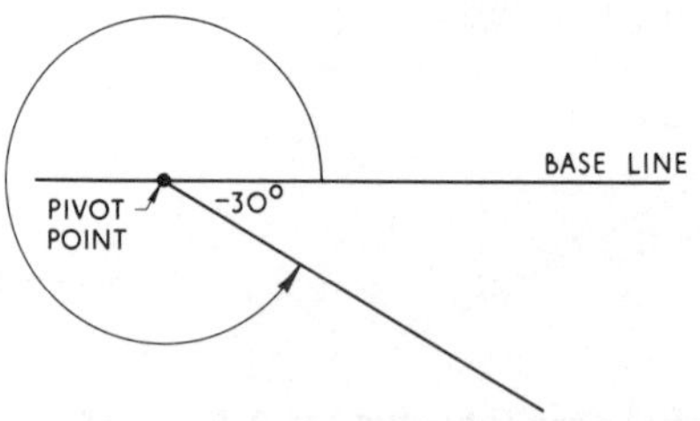

There is no need now for us to stop at 90 degrees or any other figure. We can go on around and around either way as long as we like. And, of course, if we do this we are bound to meet ourselves coming back.

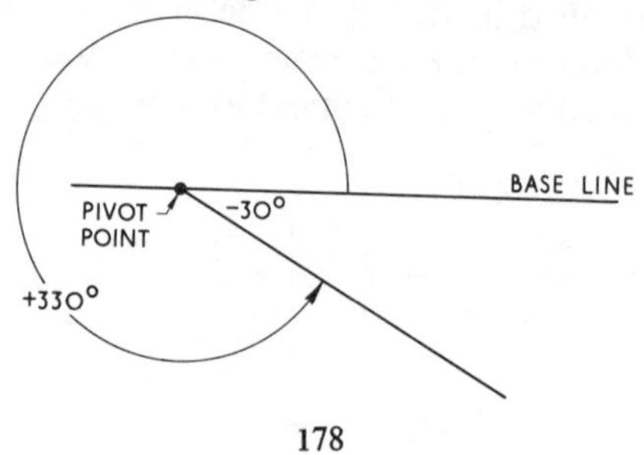

We find that, trigonometrically speaking, an angle of -30 degrees is the same as an angle of $+330$ degrees and so on. Thus if we are adding angles, we go around one way; if we are subtracting them we go around the other. With this arrangement we find that the arithmetic of measurement now applies both to the sides and to the angles of our right-angled triangles. Let us see how a test case works out, beginning with the triangle shown below.

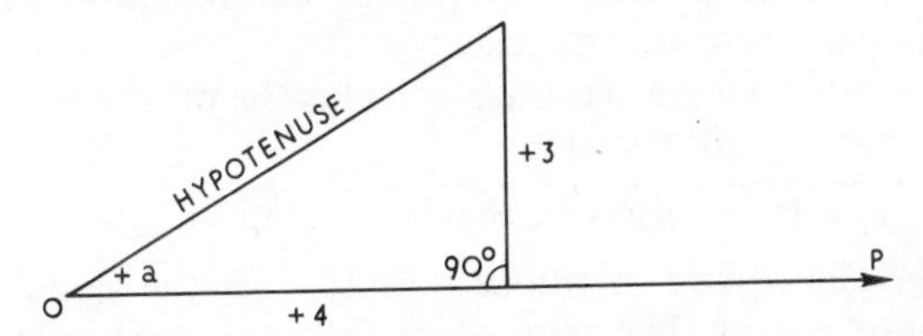

This is our old familiar right-angled triangle. In this case the hypotenuse is equal to $\sqrt{3^2+4^2} = \sqrt{25} = \pm 5$. The only difference between this and previous triangles is that now we have put a plus in front of the 3 and the 4 and angle "*a*" to show that they have a specified direction. The hypotenuse refuses to commit itself and works out to plus or minus. It couldn't do much else because it hasn't got a definable direction in relation either to the horizontal or vertical directions. What it has got is a definable slope, which in this case is ${}^{+3}/_{+4} = \frac{3}{4} = +0.75$. The slope is of course the same as the tangent ratio of angle *a*. In an ordinary triangle the sine and cosine ratios are always regarded as positive quantities. Since we have already decided that in the triangle shown the other two sides will be plus, we must make the hypotenuse plus also or else our ratios will become negative. So we have the sine of angle *a* equal to ${}^{+3}/_{+5} = +0.6$ and the cosine ratio equal to ${}^{+4}/_{+5} = +0.8$.

Now suppose we subtract 8 units of length from the plus 4 horizontal units of our triangle. The sketch below shows what happens to our triangle. The unfortunate creature is turned inside out like an umbrella in a storm. Like this:

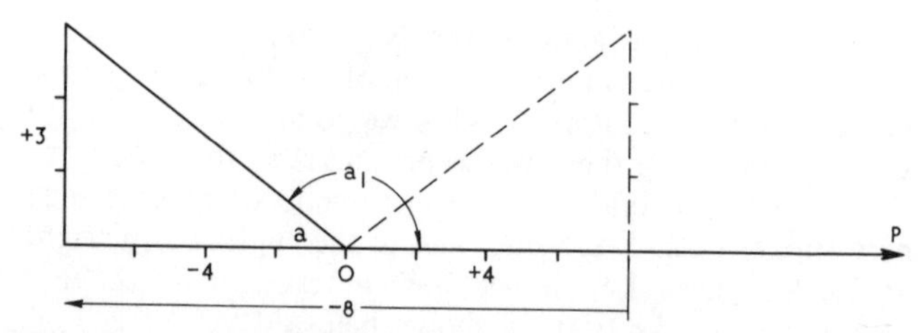

(The dotted lines show the triangle before the eight horizontal units were subtracted.)

Now our triangle has sides of plus 3 and minus 4. The hypotenuse is still the same,

$$\sqrt{+3^2+(-4)^2} = \sqrt{9+16} = \sqrt{25} = \pm 5$$

because the square of minus 4 is 16, just the same as the square of plus 4. The tangent of the angle, however, is no longer $^{+3}/_{+4}$ but is $^{+3}/_{-4}$ which is $-\frac{3}{4}$, that is, minus 0.75. If the angle was still the original angle *a* this wouldn't make sense. But when we gave direction to the lines we also gave it to the angle, and all angles are now measured by their amount of rotation from the original line marked *OP*. By this standard our angle is not the angle *a* within the new triangle but is a larger angle, which is well over 90 degrees and getting close to 180 degrees. Its actual value is 180 degrees minus angle *a*.

Suppose now we subtract 6 vertical units from the second triangle. Again our triangle is transposed, like this:

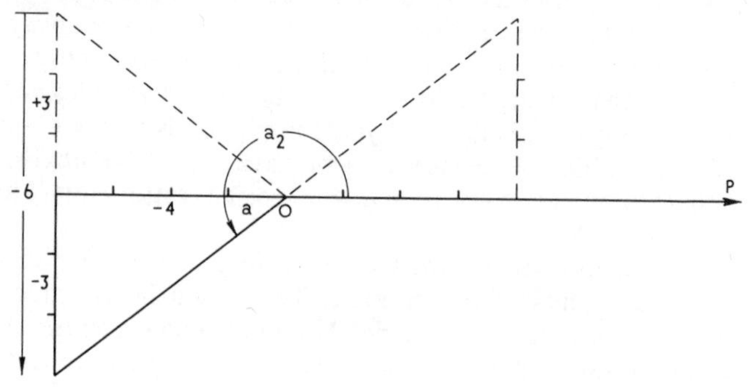

Now our two sides are both negative and the slope is $^{-3}/_{-4}$ which is $+\frac{3}{4}$ which is the same as it was at first. We can see from the sketch that this is so. The angle a_2 has now grown to 180 plus *a* degrees.

If we now add eight horizontal units to our triangle, we turn it inside out once again and finish in the fourth and last corner. Like this:

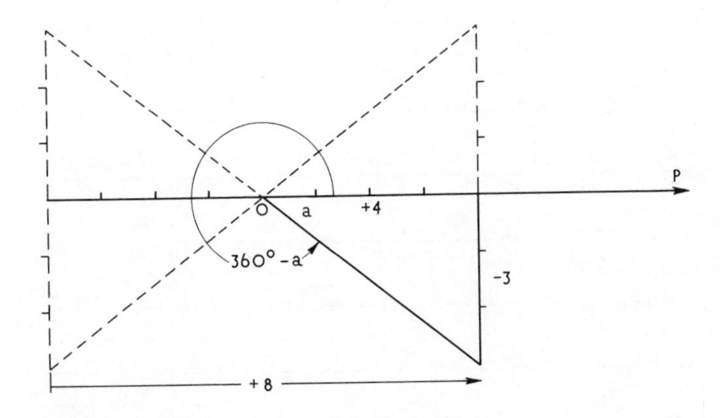

Here we have the horizontal side plus 4 again while the vertical side is still minus 3. The tangent ratio is again negative ($^{-3}/_{+4}$) which means, as we can see, that the slope has been reversed. The angle has now increased to 360 degrees minus *a*.

If we make one more move by adding six vertical units to our triangle, we will be right back where we started from, and if we like we can go around and around forever. The second time around we can call the angle either *a* degrees or 360 plus "*a*" degrees whichever we like. The results will be the same in both cases.

To give a summary, we show again the four triangles with the angles shown and the sides marked with the appropriate plus or minus direction.

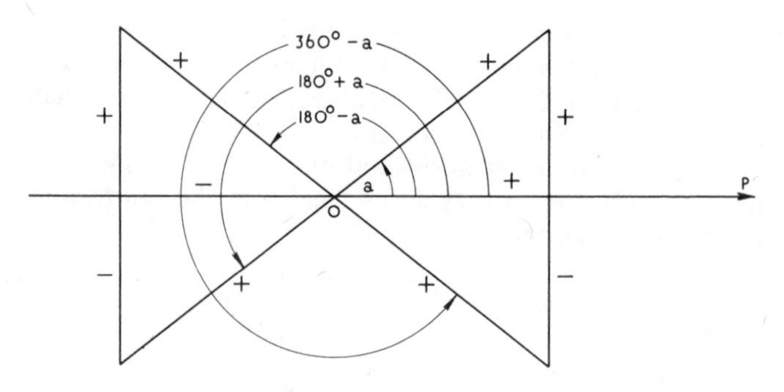

The signs of the ratios are:

Angle	Sine	Cosine	Tangent
a	$\frac{+}{+} = +$	$\frac{+}{+} = +$	$\frac{+}{+} = +$
$180-a$	$\frac{+}{+} = +$	$\frac{-}{+} = -$	$\frac{+}{-} = -$
$180+a$	$\frac{-}{+} = -$	$\frac{-}{+} = -$	$\frac{-}{-} = +$
$360-a$	$\frac{-}{+} = -$	$\frac{+}{+} = +$	$\frac{-}{+} = -$

At this stage we can pause and ask what is the purpose of putting positive and negative directions into trigonometry and of working out ratios for these fancy angles.

We have mentioned before that mathematics is a kind of language. And one of the things about a language is that it has to be constantly changing to keep up with the times. New words are invented and old words sometimes acquire almost completely new meanings. In this field of change, what shook mathematics out of its slumbers was the invention of the steam engine and all the machinery that went with it. And the heart of practically all this new machinery was an ingenious piece of mechanism known as a crank.

There is a saying that a crank is something (or someone) which will, if pushed hard enough, cause a revolution. It is certainly true that a crank (a mechanical, not a human one) was responsible for the revolution in trigonometry which we have just been discussing.

All of us are familiar with at least one form of mechanical crank, namely, the pedal mechanism on a bicycle. The essential point about it is that it provides a means of changing an up and down kind of motion into a circular one. We push *down* on the bicycle pedal and the wheels start to go around. It is when we want to find out how much downward movement we need to give for a certain amount of rotation that the new trigonometry comes to our aid. The sketch below will make the relationship clear. In order to be consistent we have shown the same reference line *OP* that we showed in the previous diagrams, and we have measured the angles from there.

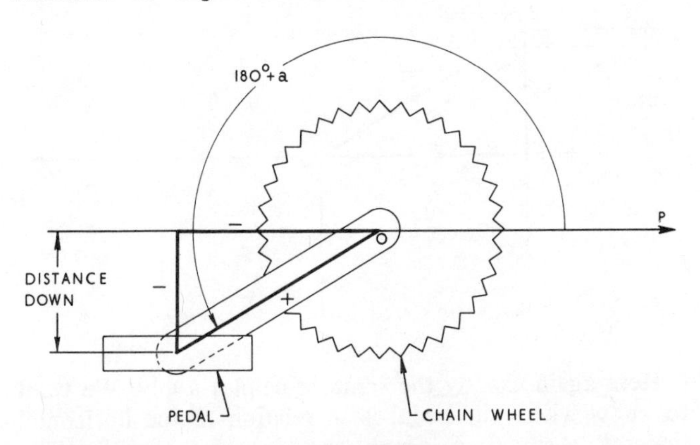

You can see the resemblance between this diagram and the diagrams of the triangles that we showed on page 180. To be consistent we have assumed that the starting point was when the pedal was pointing to the right and half way up, that is, along the reference line *OP*. Since then it has travelled 180 degrees plus angle *a*. What we want to know is the relationship between the height of the pedal center and its original position.

If we look at the diagram for a minute, we will realize that the length of the pedal arm is equivalent to the hypotenuse of our right-angled triangle, and the distance down is the opposite side. So the relationship between them is simply the distance down divided by the hypotenuse, which gives the sine ratio of the angle. And if we look back we will see that the sine ratio of an angle whose value is 180 plus *a* degrees is negative, which means the pedal is a negative distance from the base line (that is, it is below the base line not above it).

Suppose the pedal hadn't got so far around and was in the position shown below.

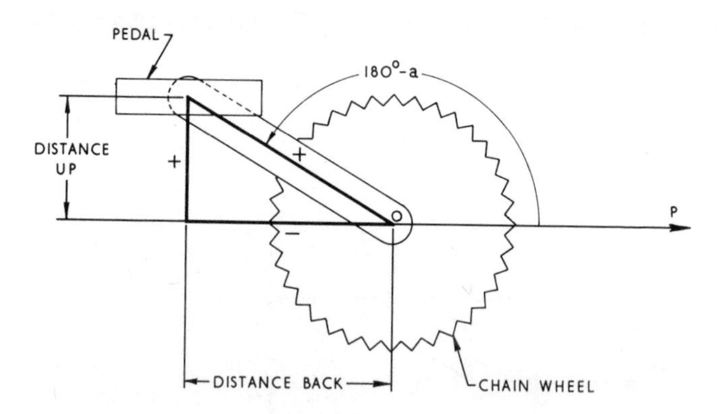

Here again exactly the same principles apply. We want to know where the pedal is in relation to the horizontal line *OP*. Again it is simply the sine of the angle 180-*a* degrees. In this case the sine is positive and numerically equal to the sine of angle *a* itself, so the pedal is above the base line. Finally, we can see that as well as the distance the pedal is above or below the base line we also get the amount by which it is forward or backward of the vertical center line. The ratio between the distance backward or forward and the length of the pedal arm is the same thing as the ratio between the adjacent side and the hypotenuse in our right-angled triangle. And this of course is the cosine ratio of the angle 180-*a*.

We have already seen that the cosine of an angle 180-*a* degrees is negative, so our answer will be a negative length, or a distance which is back from the vertical center line. From the diagram we can see this is so.

Thus, if we know the angle, we can find the exact position, up or down, backward or forward, and we can work out the angle if we know the position of the pedal.

It is hardly an exaggeration to say that without these ratios we would not be able to design any kind of machinery. Every mechanical press has a crank type of movement similar to this:

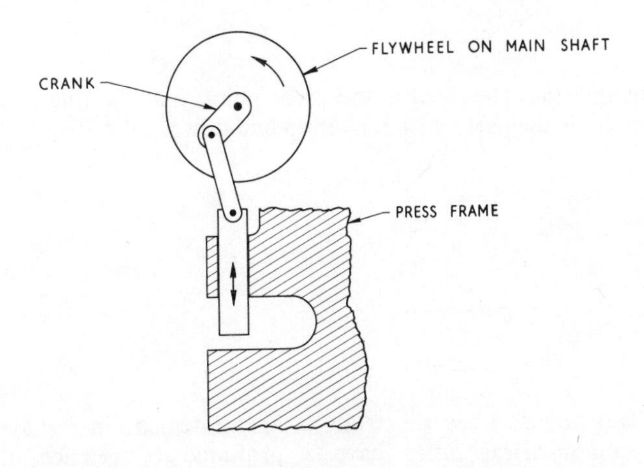

The press moves up and down as the main shaft rotates. Every steam and automobile engine works on the same principle. A piston moves up and down as the crankshaft rotates. The piston, just like the bicycle pedal, is what pushes the crankshaft round.

Even if you are fishing and you swing the rod up to lift the line out of the water, you are using the same principle.

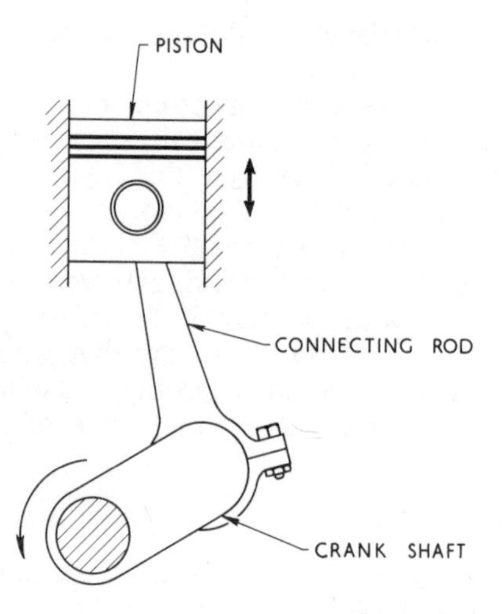

In this case the wrist is the pivot point, and the line and sinker is the part which moves up and down. Like this:

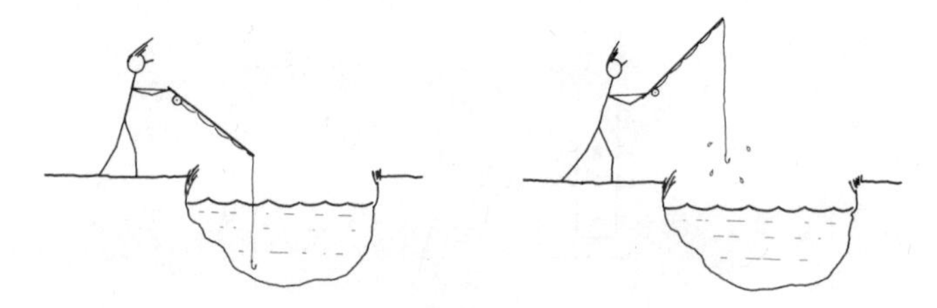

So you can see the truth of the statement that these trigonometrical ratios have a profound importance in almost every aspect of our everyday lives.

As soon as a person becomes important in life, we find hordes of newspaper photographers rushing along to photograph him. If he becomes really famous, some artist may paint his portrait. It's just the same in mathematics. As soon as a ratio becomes famous somebody tries to make a picture of it in the form of a graph. The new trigonometrical ratios were no exception. Obviously the most useful pictures would be pictures showing the relationship between the angle through which the hypotenuse had rotated and the vertical (and also horizontal) distance from the center pivot point. We can begin by making a picture showing the relationship between the amount of rotation and the vertical height. We start off by drawing an extended base line and setting a starting point for the graph like this:

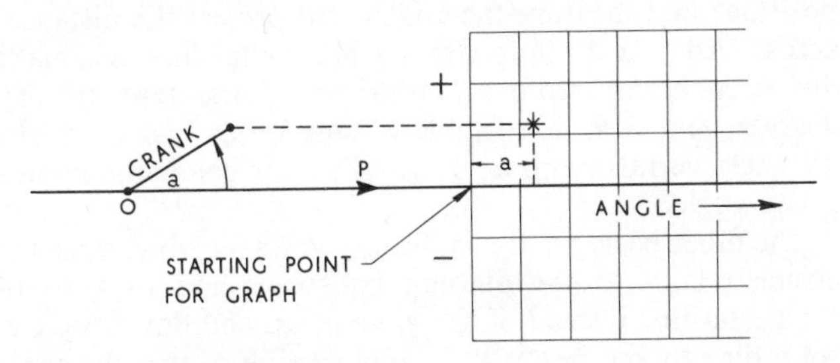

Next we imagine our crank rotating. We can see that the crank pin as it rotates will describe a circular path with the pivot point as center. To save a number of diagrams we can show, on the one sketch, the crank in a number of positions, each one corresponding to a certain amount of rotation. Like this:

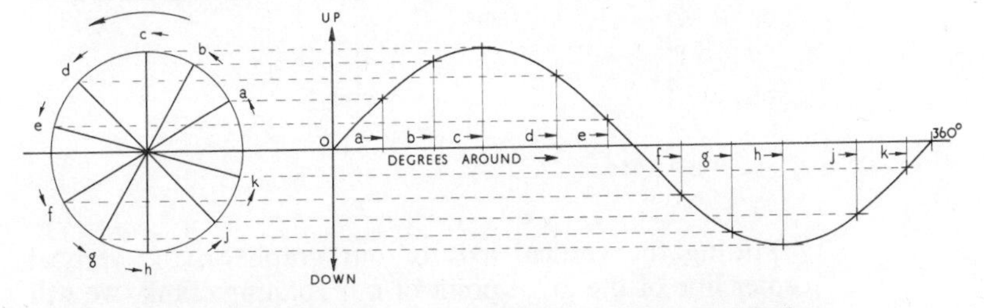

For each angle we have measured the distance of the crank pin up or down and plotted it on the vertical axis of the graph against the appropriate angle in degrees along the horizontal graph axis. And as a result we get one of the most famous and important graph shapes known to science. It is called a sine wave and for once the name is completely appropriate. (Probably it was named by practical engineers and not by mathematicians!) It has a wave-like shape—actually the ocean swell, when the sea is smooth, has this same basic shape—and it is called a sine wave because the height at any particular angle is equal to the sine of that angle. Thus, if you want to plot a sine wave, you don't have to draw a whole lot of cranks in different positions and measure the angles and project the distances across. All you do is to draw a horizontal line and make distances along representing angles from zero to 360 degrees, and then get a table of sine ratios and mark the distances vertically up or down above or below the appropriate angle.

The other basic picture is the cosine wave. Now, since the cosine relates to the amount backward and forward of the center line instead of the amount up and down, we can get it directly by simply altering our graph so that the angle is now measured along the vertical axis, like this:

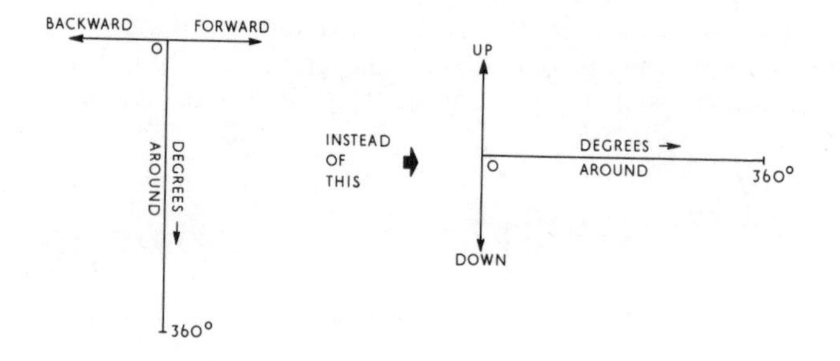

Putting the vertical axis of our graph on the vertical center line of the pivot point of our rotating crank, we will then be in a position to project the amounts backward and forward directly down on to our new graph in exactly the same way that we projected the up-and-down movements across on to our previous graph.

So here we have the cosine graph. And again if we want to plot an accurate one we don't have to do it this way. We simply get a table of cosines and mark the distances against the appropriate angles.

If we compare the graphs of the sine and cosine a very interesting fact emerges. We find that both curves are exactly the same shape. The only difference is that the sine curve starts at nothing at 0 degrees and works up to the maximum distance (or displacement) at 90 degrees. The cosine curve, on the other hand, starts with the maximum displacement at 0 degrees and goes down to nothing at 90 degrees. In other words, if we pushed the cosine curve along a "distance" of 90 degrees on the graph, it would be exactly the same as the sine curve. When two curves are exactly the same shape but are displaced in this way we say they are out of phase. In this way the sine and cosine curves are 90 degrees out of phase with each other.

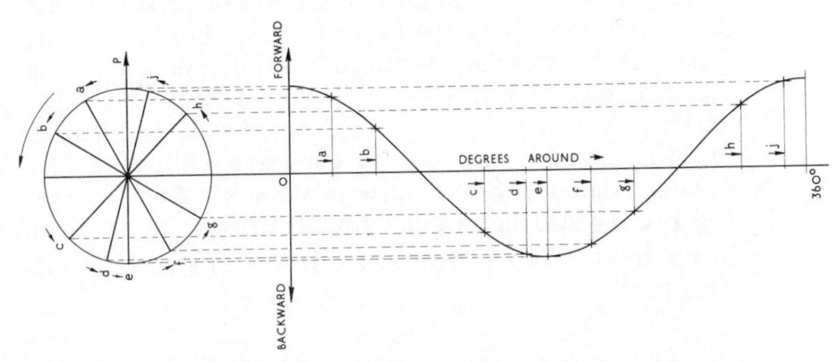

As well as mechanical engineering we find that almost the whole of radio and electrical engineering mathematics is based on sine waves. The radio waves which come from broadcasting stations are basically sine waves. The radiation from the sun, and all other kinds of light waves, are basically sine waves. Even the pendulum of a clock, or the vibrations in a musical instrument, the sounds we hear, the branch of a tree swinging in the breeze, all these things have a movement which is directly related to sine waves.

There is an old saying, "Scratch a Russian and you'll find a Tartar." This may or may not be true, but if we amend the saying to read "Scratch any kind of regular movement and you'll find a sine wave," we will, without exaggeration, be expressing one of the most profound and general truths that exist in the universe as we know it. From all this you can see how useful a mathematics that deals with the measurements of sine waves can be.

Why has nature got such a "crush" on sine waves? We will try to explain this in a later section.

Among the readers of any technical books or articles there are always a few awkward people who write in to point out that the illustration on page 874 couldn't possibly work because the thingamabob lever is longer than the gafoozalit. Or that the implications of the statement on page 196 directly contradict the implications of the statement on page 2045. Such people are invaluable. They prevent writers from sitting back and imagining that at last the perfect book or the perfect subject has been produced.

And at this stage an alert reader is quite likely to point out that, way back in Chapter 1 when talking about distances and directions, we pointed out that if we adopted the idea that distance in one direction was positive, we had to adopt the idea that distance in the other direction was negative and also the idea that distance in a direction at right angles must have the label i (that is $\sqrt{-1}$) attached to it. Like this:

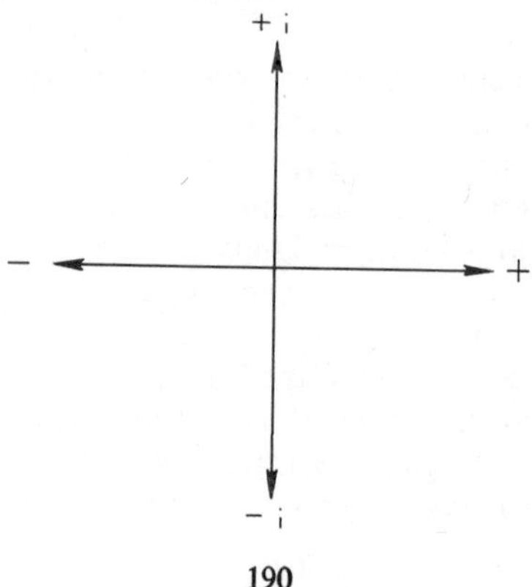

Yet in this chapter we have been talking about slopes of one in one or one in two and so on and have been drawing sketches like these:

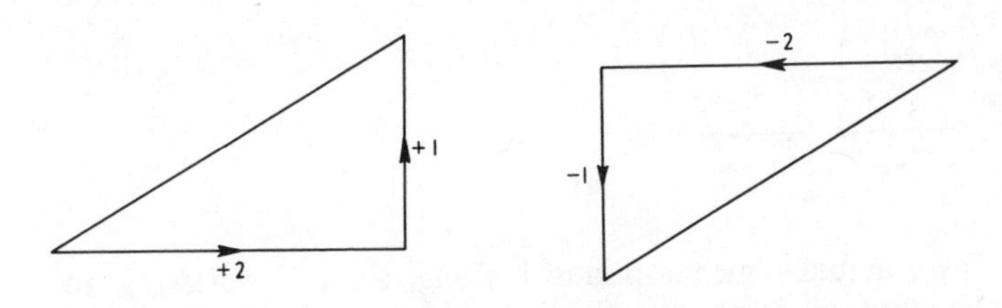

What right have we to ignore the *i* and give plain numbers to distances at right angles to each other? The answer is tied up with the fact that we have not been talking about distances and directions on maps but about slopes, and slopes are ratios. As we already know, in a ratio we can divide totally different things and get sensible results. Like dividing a hundred pounds by one foot to get the ratio of stretch to pull in a rope, or dividing miles by hours to get the ratio called speed.

The fact that miles and hours are totally different things and one can't convert one into the other doesn't matter at all. In the same way with the ratio called a slope, it doesn't matter if the distance up and the distance along are two quite different things and we can't convert one into the other. In fact, the idea of distance up (or down) being divided by the distance along (backward or forward) is inherent in the definition of slope. As pointed out before, a slope of 1/2 doesn't just mean one half (although we can use it as if it did when we are doing calculations). It actually means a slope of a steepness equivalent of one *up* for every two *forward.*

The symbol "*i*" is an operator which swings a line counterclockwise through 90 degrees.

It follows that if we talked about a slope being 1*i*/2 we would really be saying that it was a slope of one up multiplied by *i* for twice that distance along. But 1 up multiplied

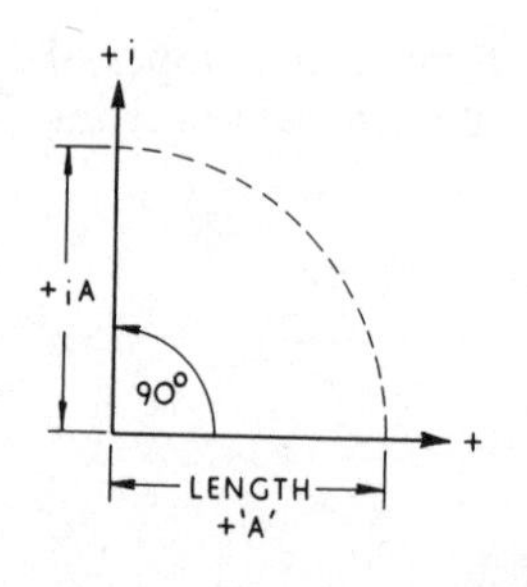

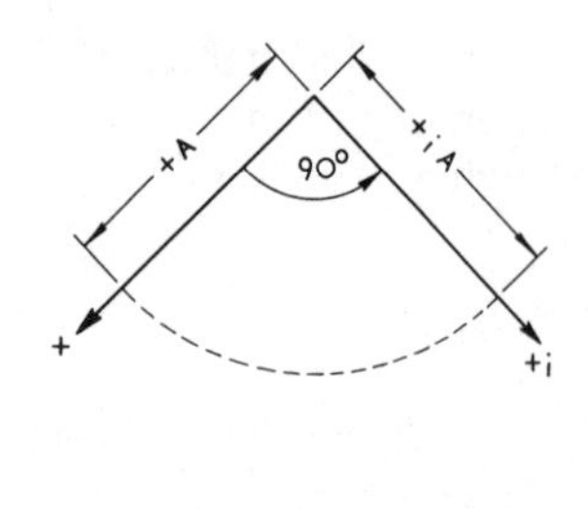

by *i* actually means minus 1 along, since if we swing an amount of 1 up counterclockwise through 90 degrees it becomes −1 along, like this:

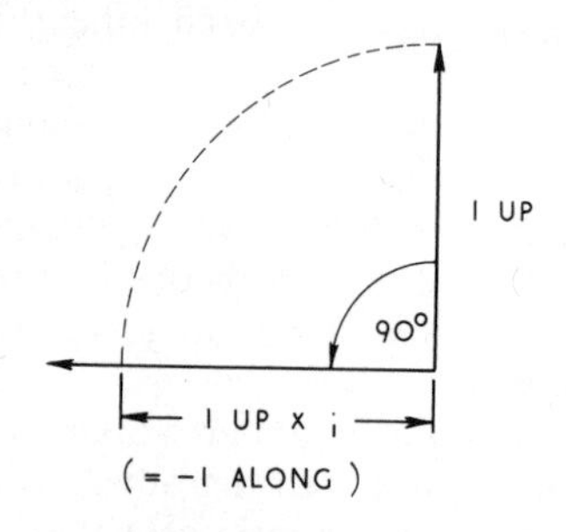

So, translated into English, a slope of $1i/2$ (or more simply $i/2$) would mean a slope of one along backward for every two along forward. Which of course is just plain rubbish.

The same argument applies to the triangles. By definition the sides concerned are at right angles to each other or they wouldn't be right-angled triangles.

Mathematics is a peculiar business. Many people sing its praises on the grounds that it represents the highest form of logical thought known to man. In practice, this isn't the case at all. Mathematics isn't so much logical as self-consistent.

You can tell mathematics that an elephant will fit comfortably into a matchbox and it will believe you without question. If somewhere among the calculations you make the statement that a kangaroo is smaller than an elephant it will accept that as well. But if you then say, on the basis of personal experience, that a kangaroo is too large to fit into a matchbox, mathematics will immediately prove conclusively that anyone who thinks a kangaroo won't fit comfortably into a matchbox is a raving lunatic.

On the whole it's safer to trust the logic of human beings than the logic of mathematics. The fact that human beings are so inconsistent at least means that they can't be consistently wrong for a hundred per cent of the time!

9. HIGHWAY TO NOWHERE

So, naturalists observe, a flea
Hath smaller fleas that on him prey;
And these have smaller still to bite 'em;
And so proceed ad infinitum. . . .

Dean Swift

There is a very old story about a frog that wanted to get to the edge of a pond a couple of feet away. He started off in fine style with a leap that took him half the distance, but then his efforts tailed off. Each leap he made was only half the distance of the previous one. In other words, he jumped a foot, then six inches, then three inches, then one and a half inches, and so on. The question is, when, if ever, did he get to the edge of the pond?

There are countless stories all from the same family tree: the story about the man who started to fly around the world, the story of the kangaroo that wanted to jump across the road, and the story of the student who set out to read a book on mathematics. Each of these problems seems to have two different answers, and you can accept either or both or neither, whichever you prefer.

The first answer, and perhaps the more obvious one, is that the frog never exactly reaches the edge of the pond. Every jump he makes only covers half the remaining distance between himself and the target; and so, however many jumps he makes, he never does anything but jump half the necessary distance.

The second answer is that the frog will eventually attain his objective if he makes an infinitely large number of jumps. From an ordinary common sense point of view, it might seem that we would always be left with an infinitely small but quite definite distance that the frog had not managed to jump. At this stage, however, a line of argument can be worked out that is plausible, full of low cunning, and also neatly shifts the onus on to the questioner. It is typical, in fact, of the kind of argument that is the delight of the mathematician. The argument goes something like this: "If you contend that the sum of all these jumps, extended to infinity, does not exactly equal the total distance to the edge, then there must be some finite distance between the frog and the edge. But whatever finite distance you name, no matter how small it is, we can prove that, after a large enough number of further hops, the frog will be far nearer to the edge than to the distance you have named. And the only place which is nearer to the edge than any place you can possibly name is exactly at the edge itself."

To misquote a popular song " . . . anywhere you can go I can go nearer. I can go anywhere nearer than you." "No you can't!" "Yes I can!" "No you can't!" And so on ad infinitum.

In actual practice nobody would worry whether the frog does ever get to the edge of the pond or whether the student ever gets to the end of the book. We could afford to ignore these particular problems were it not that basically similar problems crop up in situations of immediate practical importance in our everyday lives.

The case of the frog and the student are particular examples of an attempt to find the sum of an infinite series of numbers which have some pattern of relationship to each other. What we are actually asking in the case of the frog is whether the sum of 1 plus $\frac{1}{2}$ plus $\frac{1}{4}$ plus $\frac{1}{8}$ and so on to infinity is exactly equal to 2. The case of the student is fundamentally the same. Here we are asking if the sum of $\frac{1}{2}$ plus $\frac{1}{4}$ plus $\frac{1}{8}$ and so on to infinity are exactly equal to 1.

We can see that the two series are really the same by simply adding 1 to the second series. Then we have 1 plus the series $\frac{1}{2}$ plus $\frac{1}{4}$ plus $\frac{1}{8}$ and so on equated to 1 plus 1. It is now identical with the first series.

Putting the problem in a more general way still, the question is: does any number plus $\frac{1}{2}$ of itself plus $\frac{1}{4}$ of itself plus $\frac{1}{8}$ of itself and so on to infinity add up to exactly twice the number?

Or, to use mathematical jargon, we can ask whether, when x represents any number, the sum to infinity of $x+\frac{1}{2}x+\frac{1}{4}x$ $+\frac{1}{8}x$ . . . is exactly equal to twice x. One's first natural reaction is that the whole question is absurd. How can we possibly add the series to infinity to find the answer?

We can't, of course, add the series to infinity, but surprisingly enough we don't need to. With the aid of some ingenious fiddling we can get the answer by elementary arithmetic.

We begin by stating the problem. The sum to infinity (we will call it $S\infty$ for short, ∞ being the common math symbol for infinity) of the series is equal to all the terms added together, like this:

$$S\infty = 1 + \tfrac{1}{2} + \tfrac{1}{4} + \tfrac{1}{8} + \tfrac{1}{16} \ldots \frac{1}{\text{Infinity}}$$

Now if we halved every one of these terms we would naturally get half the sum to infinity. Take for example the sum of the first three terms. These are 1 plus $\frac{1}{2}$ plus $\frac{1}{4}$ which adds up to $1\frac{3}{4}$. If we halved each of these terms we would have $\frac{1}{2}$ plus $\frac{1}{4}$ plus $\frac{1}{8}$ which adds up to $\frac{7}{8}$ which is half of $1\frac{3}{4}$.

So we can say that if

$$S\infty = 1 + \tfrac{1}{2} + \tfrac{1}{4} + \tfrac{1}{8} \ldots \frac{1}{\text{Infinity}}$$

$$\tfrac{1}{2} \text{ of } S\infty = \tfrac{1}{2} + \tfrac{1}{4} + \tfrac{1}{8} + \tfrac{1}{16} \ldots \frac{1}{\text{Infinity}}$$

Let us write the series underneath each other, and subtract one from the other. Like this:

$$S\infty = 1 + \tfrac{1}{2} + \tfrac{1}{4} + \tfrac{1}{8} + \tfrac{1}{16} \ldots \frac{1}{\text{Infinity}}$$

$$\text{Subtract } \tfrac{1}{2} \times S\infty = \quad + \tfrac{1}{2} + \tfrac{1}{4} + \tfrac{1}{8} + \tfrac{1}{16} \ldots \frac{1}{\text{Infinity}}$$

$$= \tfrac{1}{2} \times S\infty = 1 + 0 + 0 + 0 + 0 \quad \ldots + 0$$

So by simple subtraction we find that $\frac{1}{2}$ the sum to infinity $=$ 1 and therefore the sum to infinity is exactly equal to 2! Simple, isn't it, and though it may seem to offend all reason and common sense it is actually perfectly true. It should be noted also that we don't get an answer of about 2 or almost 2 but the answer is *exactly* 2.

In spite of all this, however, we don't really dodge the dilemma mentioned at the beginning of the section. When we divided the last term of the series (that is $\frac{1}{\text{Infinity}}$) by 2 it still remained $\frac{1}{\text{Infinity}}$. So we are actually making the assertion that half of infinity, or twice infinity, or any other number of times infinity is still infinity.

Actually, it sounds a lot more reasonable if we put things the other way around and say that an infinite number of twos equals infinity and so does an infinite number of threes or ones or halves or any other quantity. So the only logical meaning we can give to $\frac{1}{\text{Infinity}}$ is that it equals zero.

Both 0 and infinity are symbols—one can hardly call them quantities—which have to be treated with a great deal of caution. Because $2 \times 0 = 0$, and $3 \times 0 = 0$, we can't cancel out the zeros and say that therefore $2 = 3$. In the same way, because $2 \times$ infinity = infinity, and $3 \times$ infinity = infinity, we can't cancel out the infinity and say that $2 = 3$.

We just have to recognize that 0 and infinity can sometimes be used as though they were quantities, but only under certain conditions.

In the last resort, the justification for using the ideas of nothing and infinity in the ways they are used lies in the fact that by doing so we get useful practical results.

For those who are still not convinced that we can get an exact answer for the sum of a series to infinity, it is worth pointing out that we take this fact for granted almost every time we do simple arithmetic!

One of the first facts we find when learning to use decimals is that some fractions, like 1/3, don't ever work out to an exact amount. If we try to convert 1/3 to a decimal we find it comes to 0.3333333 . . . and so on for ever. To indicate what is going on we put a little dot over the point three, like this, $0.\dot{3}$, and call it point three repeating. What we may not have noticed is that our point 3 repeating is actually the sum to infinity of 3/10 plus 3/100 plus 3/1000 plus 3/10000 and so on, in just the same way as the previous infinite series.

The interesting point is that here we have worked backwards, as it were, and we have found, not an amount which represents the sum of an infinite series, but an infinite series the sum of which represents an amount, namely 1/3. And this amount is not approximately 1/3, or 1/3 less some infinitely small quantity, but 1/3 precisely and exactly. We know this because we started with 1/3 exactly and precisely.

We can quite easily find lots of similar series which give a simple and exact answer if their sum to infinity is taken. 1/9 for instance is 0.111111 . . . which is 1/10 plus 1/100 plus 1/1000 plus 1/10000 and so on forever, while 1/7 is a little more complicated but basically the same. 1/7 in decimals is 0.142857 . . . and then it repeats all over again, 142857142857 . . . until the cows come home. In this case, we can see that the sum of quite a messy-looking series, namely 1/10 plus 4/100 plus 2/1,000 plus 8/10,000 plus 5/100,000 plus 7/1,000,000 plus 1/10,000,000 and so on to infinity, comes to the simple fraction 1/7.

There is of course no reason why every series when summed to infinity should give a nice simple answer. Some quite simple looking ones give an answer that is in effect another series summed to infinity, but expressed this time in decimals, which of course is a much more manageable form. We can use as much as we want for any particular degree of accuracy and ignore the remainder. The symbol π which is the ratio between the circumference and the diameter of a circle, can be expressed in the form of a series which works out to 3.14159265 and the rest. Before the days of computers people used to enjoy themselves calculating it to several hundred places, presumably in the hope that it might eventually work out. Mathematicians now claim that it will go on for ever and it has been computed, with electronic machines, to hundreds of thousands of places. The figure $3\frac{1}{7}$ however is close enough for most simple calculations, so the matter is a bit academic.

Up to now we have dealt only with series of numbers where each one is smaller than the previous one. There are series also where the numbers get larger instead of smaller. Our ordinary numbers, 1, 2, 3, 4, 5, 6, 7, and so on, form this kind of series. It is obvious that if we try to add a series like this to infinity the answer will be infinity. In the example above the last number will be the same as the number of numbers (or terms, as they are called) in the series. For instance, the fifth term is 5, the sixth term is 6, and so on. So if the series has an infinite number of terms the last one will be equal to infinity even before we add all the other terms to it. Though we can't sum this series to infinity, even the mathematicians of the ancient world discovered quite a lot about this kind of series. For instance they found rules for adding a specified number of terms. The one we have just mentioned, a series consisting of the ordinary numbers, follows a fairly simple rule. The first term, 1, is equal to itself; the first two, 1 plus 2, equal 3; the first three, 1 plus 2 plus 3, equal 6; the first four, 1 plus 2 plus 3 plus 4, equal 10, and so on (incidentally the sum of the terms give us a new series, 1, 3, 6, 10, 15, 21, etc!).

There are a whole lot of different kinds of series and anyone who wishes to delve into the business further can find all the information he wants in a standard textbook. But in passing it is worth noting a typical method by which mathematicians have worked out answers to what would be the sum of various numbers of terms in this kind of series.

A series like the example above, $1+2+3+4\ldots$ etc., does not present much difficulty at all. It is about the simplest kind we can get. In the first place, the problem of summing an infinite number of terms does not arise because, as we have just seen, the sum of an infinite number of terms is simply infinity. So the only thing we have to do is to find the answer to the sum of a definite number of terms. We could arrive at the answer by simply writing down all the terms and adding them together, but with a little thought we can find a much easier and quicker way of getting the answer.

Suppose we wanted to add together all the terms in the series $1+2+3+4+5+6+7+8+9$. It is quite easy to see that the sum will be the same if we rearrange the series by writing it backwards like this: $9+8+7+6+5+4+3+2+1$. But if we add the individual terms of the series written the usual way to the individual corresponding terms of the series written backwards we find that a very interesting fact emerges. This is shown in the addition below:

Series	$1 +2 +3 +4 +5 +6 +7 +8 +9$
Same series backwards	$9 +8 +7 +6 +5 +4 +3 +2 +1$
Sum (which is twice the sum of the series)	$10+10+10+10+10+10+10+10+10$

In other words, we find that the sum of the first and last, the second and second last, the third and third last, etc. terms of the series always add up to the same quantity and this quantity is equal to the sum of the first and the last terms of the series.

So in the example we have just given we find that twice the sum of the series is equal to 9 times 10 which is 90, and the sum of the series itself is equal to half of this which is 45. At first sight it might seem that we are behaving like the Martian who found the number of sheep by counting their legs and dividing by four. But in our case, by going what is apparently the long way around we have discovered a fact that can in future be applied to any series of this kind. From the calculation above we find that the answer we are after is equal to the sum of the first and the last (in this case 1 plus 9) multiplied by the number of terms (in this case 9) and divided by two.

So now, if we wanted to find out the sum of the series 1 plus 2 plus 3 plus . . . up to 29, we would simply add the first and last terms (29 plus 1 equals 30), multiply by the number of terms (29 times 30 equals 870) and halve it (870/2 equals 435) and we have 435 as our answer. Try working it out the hard way and you will see how useful and time-saving it is to have some simple general rule which will provide a short cut to the answer.

One might still wonder who wants to find out the sum of a series anyway. In actual fact all our arithmetic, not only recurring decimals, is based on doing just this. If we look closely at any quantity, we will find it is the sum of an irregular series of numbers—hundreds, tens, units, tenths, and so on. For instance, 195.359 is actually the sum of 100 plus 90 plus 5 plus 3/10 plus 5/100 plus 9/1000. Again, this may seem a roundabout and rather clumsy way of describing any particular number.

The advantage of this method really shows up when we have to compare two or more different numbers. Take for example the number 750 (which is 7 hundreds plus 5 tens) and 75 (which is 7 tens and 5 units) and 7.5 (which is 7 units plus 5/10), and 0.75 (which is $\frac{7}{10}$ plus $\frac{5}{100}$). Without the slightest difficulty we can immediately see that each of these numbers is ten times less than the previous one, and we can also see that a fraction like $\frac{750}{1000}$ is the same as 0.75.

The people of ancient civilizations, who did not express numbers in this way but had a separate individual name for each number, were in trouble. How many *V*'s were equal to *L*? Or perhaps it was how many *L*'s made a *V*? They might know that *V* was the name of the number that was the number of people in the family and *L* was the name of the number that was the number of years the father had lived; but unless, by previous counting, they had found *and remembered* that ten lots of *V*'s (that is *X* lots of *V*'s) added together made one *L*, they would still have had no idea of the relationship between the two numbers. The mathematicians of ancient times had to remember separately the relationship between every pair of numbers they used or else work it out by slow and painful calculation. When numbers increased beyond a few hundred the job became impossible. The modern use of a series, i.e., classifying a number as being so many thousands plus so many hundreds plus so many tens, etc., has completely changed all this. Although we may never have seen the number 8370 or the number 837 before we can see at once that the first is ten times greater than the second because we realize that while the first is the sum of 8 thousands + 3 hundreds + 7 tens, the second is 8 hundreds + 3 tens + 7 units.

If we did not express our numbers as the sum of the terms of a series it would be impossible to perform the sort of calculations that are essential in our modern civilization.

Not all series, of course, are useful. Some are useless brutes which simply make life more difficult. Take for instance the apparently simple and innocent-looking series 1 −1 +1 −1 and so on to infinity. Clearly the answer can't be 1 because there will always be a minus 1 following to cancel it out, and equally clearly it can't be 0 because there will always be a plus 1 following to make it 1 again.

Mathematicians about the time of Newton came across this series and were very puzzled about it. The most puzzling thing was that one could get a series like this through simple and perfectly legitimate division.

When we normally do a division we often have a remainder left over. This remainder is less than the amount we are dividing by, but there is no law saying it has to be so. For instance, if we divide 9 by 2 we normally get 4 plus a remainder of 1. But we would be just as correct mathematically to say that 9 divided by 2 gave 2 plus a remainder of 5, or 3 plus a remainder of 3. For two times two plus five makes nine just as does three times two plus three, or four times two plus one. Furthermore, there is no reason why, if we accept the idea of negative numbers, we shouldn't go further still. Since 9 is equal to 10−1 we can say that 9 divided by 2 gives us 5 with a remainder of minus 1, just as truly as we can say that it is equal to 4 with a remainder of plus 1. We actually do this kind of thing with logarithms.

Let us apply this reasoning to an extremely simple case, that of 1 divided by 2. With ordinary division we get the answer 0, plus 1 remainder, but if we say that 1 is equal to $2-1$ we can say that $2-1$ divided by 2 gives us 1 plus a remainder of -1. Then we can turn around and say that -1 is equal to -2 plus 1 and divide this by two again. This time we find that -2 plus 1 divided by 2 gives us -1 with a remainder of plus 1. And so we go on to find that 1 divided by 2 can give us the answer $1-1+1-1+1-1+1-1\ldots$ and so on to infinity. Which is certainly a peculiar answer!

The fact appears to be that a series such as this does not give any answer when we try to sum it to infinity. It does not give us an exact real answer like the recurring decimal series, or the series 1 plus $\frac{1}{2}$ plus $\frac{1}{4}$ etc. . . . , and it does not give us infinity as an answer as do the series 1 plus 2 plus 3 plus 4 and so on, and others, when they are extended to infinity. On the other hand if we add a real number of the terms of this series, we will find, no matter how many terms we take, that the answer is either 0 with a remainder of plus $\frac{1}{2}$, or else it is 1 with a remainder of $-\frac{1}{2}$. Which is quite reasonable because 0 plus $\frac{1}{2}$ and $1-\frac{1}{2}$ are both different ways of writing plus $\frac{1}{2}$.

It is interesting to speculate on possible meanings one can attach to the fact that we can't get any kind of answer for $1-1+1-1\ldots$ and so on summed to infinity, while with other series we can not only get an answer like "infinity" (which at least is some kind of answer) but can do even better and get an answer like 1/3, or 1/9, or 1, or 2, which are all definite as well as real numbers. Does it mean that for these latter series there is such a place (or quantity or distance, call it what you will) as infinity, while, on the other hand, there is no such place as far as a series like $1-1+1-1+1-1\ldots$ is concerned?

If one has the time one can have a lot of fun trying to work out the answers to such questions.

We can sum up by saying that roughly there are three

types of series. Firstly, there are those in which the terms go on increasing without limit and where the sum of the terms of such a series to infinity is infinity itself. Secondly, there are series like $1-1+1-1+1-1$ ... which swing backwards and forwards (or oscillate) for ever. These just won't give any answer if we try to sum them to infinity. Finally, there are the series that are perhaps of the greatest practical use, those in which the terms get progressively smaller and smaller, and that can be summed to infinity to give real, definite, and often extremely useful quantities.

Two of these quantities are so important that they are worth a special mention of their own. The symbol π, to which we have already referred, is the symbol which represents the relationship between the diameter and the circumference of a circle. The second is the symbol "e" about which we will say more later.

The problem of finding the relationship between the diameter and the circumference of a circle is one which has been of practical interest ever since man began to use wheels and rollers for transport, and circles for drawing and geometry. The very earliest people who took an interest in these things simply measured the diameter and circumference as best they could and found that the circumference was about three times the diameter. Gradually the estimate became closer and closer, but there are limits to the accuracy with which one can make physical measurements. It was then that the use of a kind of series was developed. The following sketches show the general way in which the problem can be tackled. The first sketch shows a circle with a square drawn inside it and another drawn outside.

If we look at the larger square we see that each side of the square is equal to the diameter of the circle. . . . Therefore the distance around the sides of this square is four times the diameter of the circle. If we think a little we will see that the total distance around the edges of the square must be greater than the distance around the circumference of the circle because the circumference takes "short cuts" inside the boundaries of the square. So, since the distance around

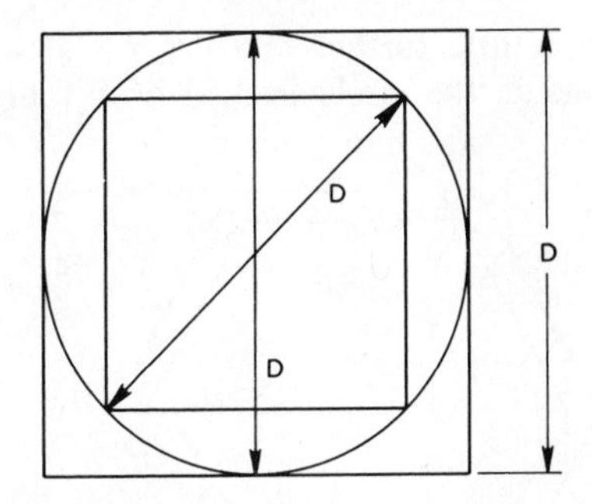

the edges of the square is equal to four times the diameter, the distance around the edge of the circle must be less than this. Now if we look at the smaller square which is drawn entirely inside the circle we will see that its *diagonal* is equal to the diameter of the circle. This diagonal divides the square into two right-angled triangles with two sides equal and with angles of 45 degrees. As we have seen before, the ratio of the sides of this kind of triangle are like this:

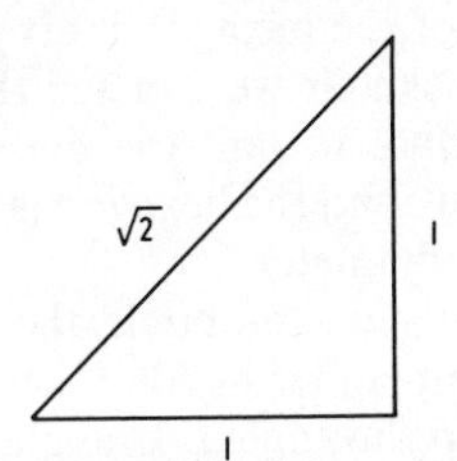

So if the $\sqrt{2}$ side is equal to the diameter of the circle then the other two sides will be in proportion to the diameter as 1 is to $\sqrt{2}$. In other words they will be $1/\sqrt{2}$ times the diameter. Since $\sqrt{2}$ works out to about 1.414 we find that $1/\sqrt{2}$ is equal to about 0.707. So the sides of the inner square are equal to 0.707 and the distance around the edge of the square—four times this—is equal to 2.828. So we can see that the circumference of the circle is larger than 2.828 times the diameter and smaller than 4 times the diameter.

We can go a little further and draw a six-sided figure inside and outside the circle instead of a four-sided one. Like this:

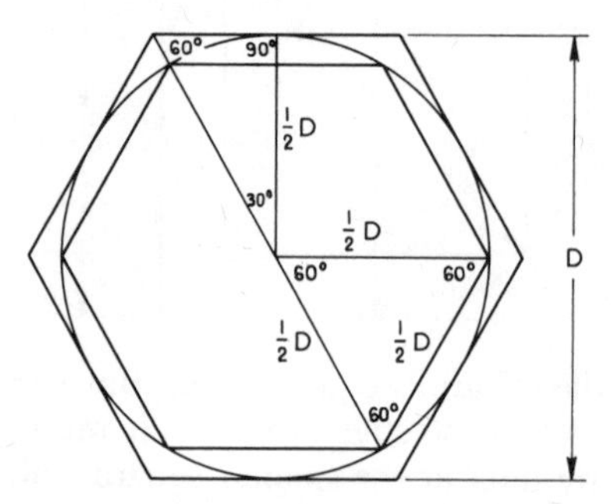

The six-sided figure is called a hexagon and we can see that in this case there are six equal angles at the center and therefore each will be a sixth of 360 degrees, or 60 degrees. It can be shown that the triangles have all their angles equal and the three sides are equal also. So the distance around the borders of the hexagon is six times the length of one side. From the sketch we can see that the side of the inside hexagon is equal to half the diameter of the circle, so that the distance around the hexagon is $6 \times \frac{1}{2}$ the diameter which is 3 times the diameter.

In the hexagon outside the circle the length of the side can be found by remembering that here we can make a pair of 30-degree right-angled triangles. As we saw in Chapter 8 the ratio of the sides is as shown below:

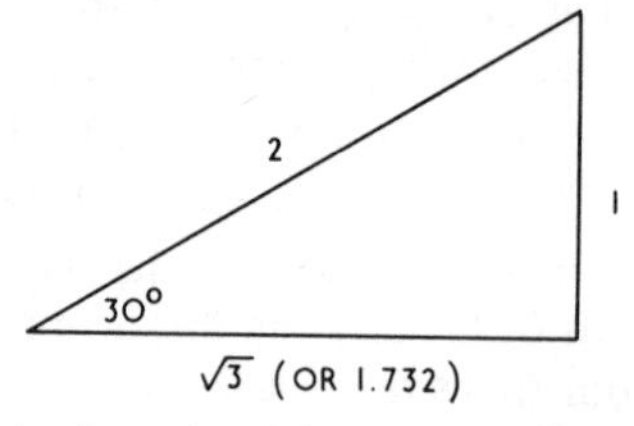

In this case the length of one side of the hexagon will be larger than half the diameter of the circle in the proportion that 2 is to 1.732 or in other words about 1.1547 times. Since the distance around the hexagon is six times this amount, the total distance around the hexagon will be $6 \times 1.1547 \times \frac{1}{2}$ of the diameter of the circle. Which works out to 3.4641 times the diameter.

Thus by using a hexagon we find that the circumference of the circle must be more than 3 times the diameter and less than 3.4641 times the diameter. With the square we were only able to say that it lay between 2.828 times the diameter and 4 times the diameter.

We can now see a pattern emerging. If we draw an eight-sided figure inside and outside the circle we will get a smaller margin still between the maximum and minimum possible limits of the ratio of the circumference to the diameter. And the sum of the ten sides of a ten-sided figure will give us much better results, while the sum of the hundred sides of a hundred-sided figure will give still better results. By this method the value for π has been calculated to a greater degree of accuracy than would ever be required, apart perhaps from calculations in astronomy. Even to get the circumference of the earth to within a mile or so we would not need an accuracy greater than five places of decimals.

But since the time of Newton, when infinite series have been used, π has been evaluated without electronic computors to some seven hundred places of decimals. To get this kind of accuracy with the method outlined would in practice be difficult, but it is possible to evolve an infinite series, of the type described at the beginning of the section, where the terms get progressively smaller and smaller. The sum of these series tends to π just as the series at the beginning of the section tended towards a whole number. The trouble is that π is not a simple number, and what its numerical value actually is probably won't be known until one of these series is literally summed to infinity. Meantime a few hundred places should be good enough for anyone. It may seem strange that π has such an awkward value but in point of fact it would be more of a coincidence if it did not. We are accustomed to "round figures" for our answers, but where natural relationships are involved it would be far more reasonable to expect the winning tickets in a lottery to consist only of numbers like 100,000, 200,000, etc., than to expect π to come out to an even value.

Another quantity that is calculated by using the sum of a series is the number known as "e." It is called a "natural number" and its numerical value is 2.71828 . . . and the rest. Like π it appears to go on forever. It seems impossible that this odd-looking number should have anything natural about it, but in fact it is a number that is lurking behind every example of natural development and growth.

Suppose we have a pound of some living tissue, perhaps like the chicken embryo which, under the right conditions, can be kept alive and growing for years. Or we can take some material that is growing through chemical reactions. Suppose we were told that all this material created an equal amount of new material over a certain period and the newly created material created still more at the same rate. If we started with one pound of the material, how much could we expect to have at the end of the period?

The answer might seem so obvious as not to be worth thinking about. At the end of the period there would be two pounds.

Actually, surprisingly enough, there wouldn't. There would be quite a lot more. If we think about the question a little more carefully, we will realize that we have overlooked the fact that, as well as the original material creating new material, we have also new material which starts creating new material itself as soon as it is created. If we think only of the original material we will think in terms of a steady and continuous growth. But if we also take into account the growth of the newly created material we will realize that at the end of the period, if there is twice as much material, the *actual amount* of creation of new material will have to be *twice as great* as it was at the beginning of the period, if the *rate of growth* is to remain constant.

This puts rather a different complexion on matters, but we are still a long way from knowing exactly how much more material will have been produced by the new material. One way we can tackle the problem is to see what is happening at a time between the beginning and the end of the period.

Suppose we take a time half way through. The original material will have increased to 1.5 pounds. Now for the second half of the period there will not be 1 pound but $1\frac{1}{2}$ pounds of material, and by the end of the period this will have increased by 50% which is $\frac{3}{4}$ pound. So the total will be not 2, but $1\frac{1}{2}$ plus $\frac{3}{4}$ which is $2\frac{1}{4}$ pounds.

Let us carry this division still further. When 25% of the period has passed, the original material will have increased by 25%, that is, there will be $1\frac{1}{4}$ pounds of material. During the next quarter of the period this $1\frac{1}{4}$ pounds will have increased by 25%, which is $\frac{5}{16}$ of a pound, making the total for the half period $1\frac{9}{16}$ pounds. For the third quarter this $1\frac{9}{16}$ pounds will have increased by 25%, which is an increase of $\frac{25}{64}$ pounds, bringing the total to $1\frac{61}{64}$ pounds. For the last quarter this amount will have increased by 25% which is $\frac{125}{256}$. So now the total amount at the end of the period is not 2 pounds but $2\frac{113}{256}$ or, to put it in decimals, 2.441 pounds.

By now it should be clear where our natural number "*e*," which is 2.71828 . . ., etc., comes in. As we divide the period into smaller parts we take more and more account of the increase due to the newly grown material itself. In actual fact, however, the newly grown material doesn't waste any time, but begins to cause further increase itself as soon as it is created. So if we are going to find the exact amount of material at the end of the period we would have to divide the period up into an infinite number of infinitely small periods and add them all together.

Surprisingly enough, mathematics is able to do just that. With the help of some rather involved mathematics which can be found in any standard textbook, it eventually comes out with the statement that *e* is equal to

$$1+1+\frac{1}{2}+\frac{1}{6}+\frac{1}{24}+\frac{1}{120}+\frac{1}{720}+\frac{1}{5040}$$

. . . and so on to infinity.

For interest, let us add the first five terms and see how close they come to our quantity "*e*." The sum works out as follows.

1st term	=	1	=	1
2nd term	=	1	=	1
3rd term	=	$\frac{1}{2}$	=	0.5
4th term	=	$\frac{1}{6}$	=	0.1667
5th term	=	$\frac{1}{24}$	=	0.042
	Total		=	2.7087

This is very near indeed. We see that the size of the terms drops off very sharply, and after a few terms the additional amount is too slight to be of practical importance. In fact we can see that all the remaining terms from the fifth on to infinity only make less than 0.01 of a difference to the answer.

Apart from its application to the way in which living things grow, "*e*" also has an application to the way interest grows if you are unfortunate enough to be borrowing on short-term compound interest. Many people don't realize that an interest rate of 50% for six months is not at all the same thing as a rate of 100% for a year, if the money is borrowed in both cases for a *whole* year. At 100% per year one has, after a year, to pay back twice the money one borrowed. But if one borrowed at a rate of 50% for six months, one would have to pay back $1\frac{1}{2}$ times the amount borrowed at the end of six months, or else for the remaining six months pay 50% on the original amount and on the interest as well. Thus for the second six months one would be paying 50% on $1\frac{1}{2}$ times the original amount, which is $\frac{3}{4}$. So at the end of the year, one would owe not twice the original amount but $2\frac{1}{4}$ times the original amount. And of course a rate of 25% for three months would be worse still. So if you are trying to borrow money at compound interest, as well as trying to get the lowest interest rate, make sure the loan is for as long as possible.

This kind of growth, where the growth of the growth has to be taken into account, is called an exponential rate of growth, and the series we have just mentioned is called an exponential series. We will meet "*e*" again in the next chapter, but in the meantime if anyone asks you how your garden is growing you can casually remark, "Oh, it's growing exponentially, of course. Just the same as it always does." Little remarks like this do such a lot toward increasing one's popularity with friends and neighbors!

10. CALCULUS AND CONTRADICTIONS

Nature, and Nature's laws, lay hid in night,
God said, Let Newton be! and all was light.
Alexander Pope

It did not last; the Devil, howling Ho;
Let Einstein be! restored the status quo.

Around 450 B.C. to 400 B.C., in what is now known as Italy, lived a philosopher who came to be known as Zeno of Elea. Actually we know very little about him and none of his writings have survived. This is hardly surprising for he seems to have been one of those nasty people who make things difficult for the established experts by putting forward questions and arguments that they couldn't answer.

As every expert knows, the only way to deal with such people is to pretend they don't exist and hope that nobody will notice them.

In Zeno's case, these tactics would probably have been completely successful, if the well-known Greek philosopher Aristotle had not imagined that he could refute some of Zeno's arguments. Full of enthusiasm he gave a fairly full account of what Zeno had said and then proceeded to refute it. The final result, however, has turned out to be the opposite of what Aristotle intended. Zeno's arguments have been remembered while Aristotle's arguments against them have been largely ignored and disregarded.

It appears that Zeno believed the universe to be a coherent and harmonious unity. When, therefore, he found that motion seemed to be self-contradictory he decided that motion was not a reality but something which existed only in man's imagination. One example he gave of the contradictions inherent in motion was that of an arrow flying through the air. Now, unless the arrow is in two places at once, it must, at any particular time, be in one place and in one place only. But if it is in one place and one place only at any particular time it can't be in motion. So if the arrow is moving it can't exist at any place, and if it exists at any particular place it can't be moving.

Another example given by Zeno was that of Achilles, who was the fastest runner of all legendary history, and the tortoise. Zeno pictured them having a race, with the tortoise starting slightly ahead of Achilles. By the time Achilles got to the place where the tortoise had started from, the tortoise had got a little bit ahead. By the time Achilles got to this second spot, the tortoise had gone a tiny bit further still. You can see the gist of the argument. Each time Achilles reached the place the tortoise had just left, the tortoise had moved a tiny bit further on. One could continue in this strain indefinitely without Achilles ever being able to pass the tortoise.

It is easy, if one doesn't give the matter much thought, to brush this off as childish nonsense. Of course the arrow moves and of course Achilles passes the tortoise. How? Well he does, that's all! But this kind of argument does nothing to answer Zeno's contention that, since the explanation is inherently contradictory, we are suffering from illusions and merely imagine we are seeing these things.

Aristotle was not particularly successful in answering Zeno. He contended that since time and distance were continuous there was no such thing as an instant or point in time; the same thing applied to space; and a line, for instance, was continuous and so could not be made up of points.

But this did not answer Zeno's contention that over any length or period of time, no matter how small, the arrow would have to be in an infinitely great number of different places and therefore couldn't be said to have an existence in any particular place.

In spite of these differences of opinion there was one point on which both Zeno and Aristotle apparently agreed. They both took it for granted that a contradiction was an absurdity and couldn't exist.

This of course is quite a natural and understandable attitude. If we see a magician appear to take a rabbit out of an empty hat, we conclude there is something phoney. Either the hat wasn't empty or the rabbit came from somewhere else.

In real life, however, the situation is not quite so simple.

There is a story about two men who were travelling to a village which lay behind a mountain. Both had made one trip years before but they had no maps and no clear idea as to where the village lay. The only thing each man remembered was some details of the road. All went well until they reached the foot of the mountain. Here the road forked, one fork went to the left and the other to the right. When they reached the fork a heated argument broke out.

"Do you think I'm a half-wit?" shouted the first man angrily. "I came here five years ago and I distinctly remember turning to the right."

"Rubbish!" answered the other man hotly. "I came here with some friends only two years ago and I know we turned to the left."

Finally, being unable to settle the argument, the two men set off in opposite directions, each convinced that he was right and the other was a fool. It was only when they both arrived at the village at exactly the same time that they realized the "two roads" were actually a single ring road which went right around the base of the mountain.

There is a story in a similar vein about two men who had completely contradictory ideas about the color of a particular automobile. At first each was sure that the other was color-blind, but when each man produced a color photograph of the car in question, and the color in each photograph was different, they decided to look into things more closely. They finally discovered that the first man had seen the car only during the daytime and the other had seen it only at night under a mercury-vapor lamp.

The point about these stories is that two completely contradictory viewpoints were both perfectly correct as far as they went. The road to the right led to the village just as surely as did the road which started off in exactly the opposite direction. The camera which photographed the car under the mercury-vapor lamp was not suffering from deficient eyesight or a hangover. The actual color or, to be even more precise, the actual wavelength of the light reflected by the car was in both cases recorded with perfect accuracy. Under the different conditions, the car was actually a different color. And while we may say that daylight is the *normal* condition under which we look at the color of an automobile we cannot say that the daylight color is "right" and the artificial-light color is "wrong," or that it doesn't exist. It existed enough for a camera to photograph it!

Most religions and moralities "which have guided mankind" present history as though it were a clear-cut struggle between the "goodies" and the "baddies"; between the saints and the sinners; and between things that are right and things that are wrong. As our knowledge increases we are beginning to realize that very often the struggle is not between the goodies and the baddies, but between a pair of goodies who are about fifty-per-cent bad (or, if you are a pessimist, between a pair of baddies who may be about fifty-per-cent good). The winner, after discarding about half of his own ideas which were impractical, and adopting about half of his late opponent's ideas which were practical, usually manages to create a state of affairs that is a slight improvement on the previous one.

So when we come up against contradictions, either in mathematics or in life, we do not need to get too downhearted, or feel like giving up. The first thing is to make sure that the contradictions are actually there. If for instance you measure the kitchen table and find it is six feet long, and someone else measures it and finds it is only five feet long, it is wiser to check your measurements than to jump to the conclusion that you have discovered some inherent contradictions in mankind's normal methods of measuring the length of kitchen tables.

Apart from this kind of thing, we find quite frequently (too frequently for many people) that there do exist real and valid contradictions which all the re-measuring and re-checking will fail to remove. It is at this stage that we should accept the validity of the contradictions and try to find the cause. It is not much use telling ourselves that the contradictions don't "really" exist, that the roads did not "really" go in opposite directions, and the car was not "really" two different colors. For when we "really" get down to fundamentals we find that the only reality that can exist for us is what we make of the impressions we gain through our senses. And if, despite all our efforts to organize things otherwise, we find that the information we get through our senses tells us that these contradictions exist, then we must accept them as real.

To accept a contradiction as real and valid does not mean that we treat it as a universal and unchanging truth. The recognition of contradictions merely means that we admit there are facts that, in the light of our existing knowledge, are incompatible with each other, and therefore don't make sense.

There may be an "ultimate reality" in which there are no contradictions, but since we have finite minds, we are faced with the contradiction that it is only after an infinite period of development that we will be able to understand this ultimate reality anyway. Meantime, if we do run into inescapable contradictions, we might as well learn to live with them.

All this may be very interesting, but why has it been introduced in a chapter that is supposed to be dealing with calculus?

The reason is that a large number of people who are trying to find what mathematics is about have painfully worked their way through arithmetic and algebra and geometry and trigonometry, only to give up in despair when they come in contact with differential and integral calculus. What they often fail to realize is that as well as a problem in mathematics they are up against primarily a problem in life itself. Immediately we begin to think carefully about objects in motion, we pass from a superficially simple state of affairs into a morass of contradictions. We find ourselves where Zeno found himself two and a half thousand years ago with the contradiction that if a thing is moving it doesn't exist anywhere and if it exists anywhere it doesn't move.

Let us see how this works out in connection with the apparently simple concept of speed.

Speed, as we know, is a ratio. We have already discussed ratios in Chapters 5 and 7 and have described them as a peculiar kind of division. We will now see that this division is even more peculiar than might have been imagined.

In Chapter 7 we showed a number of graphs. Among them was one on page 132 which represented a picture of a speed of ten miles per hour. It looked like this:

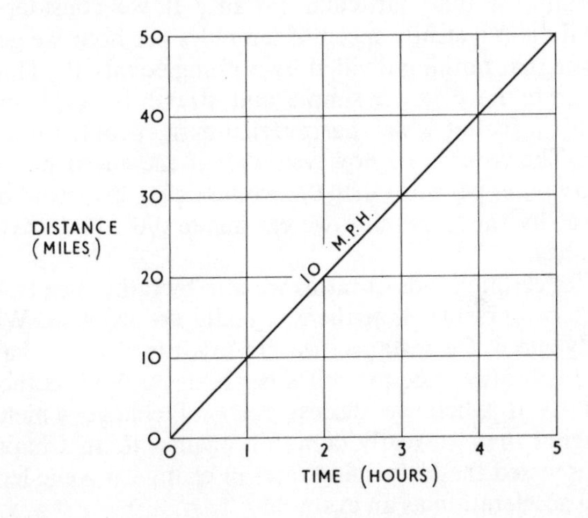

We can see that a speed of ten miles per hour means that after one hour we have gone ten miles, after two hours we have gone twenty miles, and so on. We just take the total distance and divide it by the total time and there is our ratio, speed, in miles per hour. Simple, isn't it?

We don't even have to go for a full hour. A speed of ten miles per hour doesn't mean one must have actually travelled ten miles. If you go five miles in half an hour it is still a speed of ten miles per hour because the ratio, five miles divided by half an hour, still equals ten just as does the ratio of thirty miles divided by three hours.

By exactly the same reasoning we can see that $2\frac{1}{2}$ miles divided by $\frac{1}{4}$ hour also gives a speed of 10 miles per hour as does 1 mile divided by 1/10 of an hour and so on. The whole question is, where are we going to stop?

If we continue long enough we will get down to an instant or point in time at which, of course, the car wouldn't have any time to travel any distance at all. It would be "frozen" like a still from a movie. In this case our ratio would be no distance divided by no time.

The sixty-four-dollar question is, at what speed is the car travelling at that particular instant? If we consider it is travelling at a steady speed of ten miles per hour we get the answer that nothing divided by nothing equals 10. Thus we appear to have got a simple and straightforward answer for 0/0, a matter which has puzzled many people for a long time. The trouble is, however, that if the speed had been twenty miles per hour then 0/0 would equal 20 instead of 10. In fact by this reasoning we can make 0/0 equal anything we fancy.

With certain kinds of ratios we can, by rather fast talking, convince ourselves that there is really no problem. With a steady speed, for instance, we can say it is obvious that the speed will always be ten miles per hour and that is the end of it. It is when we discuss rates of change which are changing that we really come up against it. In Chapter 7 we discussed these changing rates of change at some length, using acceleration as an example.

The graph representing an acceleration of 10 miles per hour is reproduced on the next page.

Here we have a different situation. The speed in miles per hour is not constant. It is changing all the time—every instant, in fact. And under these circumstances we can no longer talk our way out of the problem by quoting average speeds, or saying the speed must be the same as it was before and after the instant in question. Here we are faced with the fact that since acceleration *is* a continuous change of speed, it is impossible for the speed at any instant to be the same as the speed at any other instant, and there is no speed *except* the speed at some particular instant. So when we are dealing with acceleration we find the ratio 0 miles/0 hours has not one but an infinite number of quite definite answers depending on where the particular instant represented by the 0 miles/0 hours was taken.

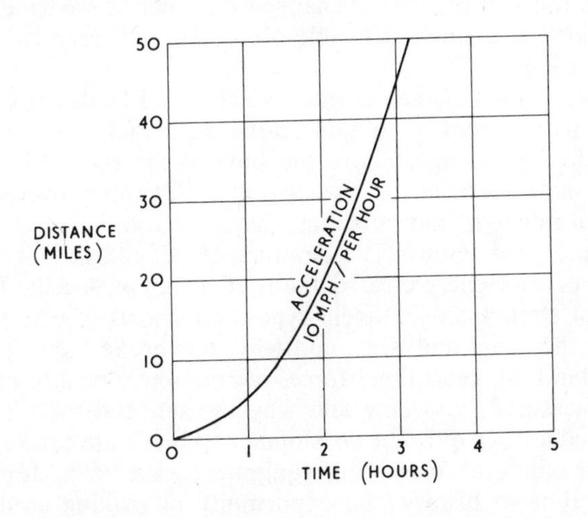

This kind of inherent contradiction is not confined to our ideas about moving objects; it is to be found in all our ideas of growth and development as well as in physical movement. In Chapter 7 a brief reference was made to the difficulty in defining ideas such as "change" and "time." We can now begin to see that there is more in these things also than might appear at a casual glance. If we think of any kind of change, from a plant growing to a piece of rock weathering away, we are faced with the fact that, apart from one single instant of time, *the* plant or *the* rock doesn't exist, because by the next instant it has changed into something different. Normally these changes are so slight and so slow that we can in practice regard it as the same plant or the same rock for a long period. But unless we are prepared to contend that an oak tree is the same thing as an acorn, or a full-grown bird is the same thing as an egg, we are forced to the conclusion that we are not dealing with one object, but a countless succession of objects, each one different from the previous one. And again, if we consider the object "frozen" at an instant of time, we have the object but have lost the idea of growth or change, while if we think of the idea of change over a period we have lost the idea of a single definable object. It is all very sad and confusing.

The branch of mathematics which could be described as the mathematics of motion really began only some three hundred years ago, about the time of Sir Isaac Newton. The mathematicians of that period didn't worry overmuch about inherent contradictions, they were too busy trying to get practical results. The development of industrialization and of machinery created the need to know specific facts about strengths of different types and shapes of materials, how they bent and when and why they broke, how speed, acceleration, centrifugal force, inertia, and pressure could be measured; and how and why the sun and the planets moved as they did. The development of accurate clocks and other kinds of measuring equipment gave man, for the first time in history, the opportunity of making accurate and reliable measurements of these things.

Most of all, the investigators of Sir Isaac Newton's day were trying to find general laws governing these happenings, so that they wouldn't have to solve each individual problem separately by a lengthy process of trial and error.

When we get down to the actual practical problem of measuring motion we find the most obvious and easiest way is not to try to measure something called "motion" directly, but to measure the amount of time the object has been travelling and the amount of distance it has travelled in that time. All that is needed for making such measurements is an accurate tape-measure and an accurate clock, and that is where Newton and his contemporaries began. Having made these measurements they were then able to draw graphs which presented a picture of the particular kind of change that was taking place.

In the examples of graphs of speed we have usually shown both the time and the distance starting from zero, like this:

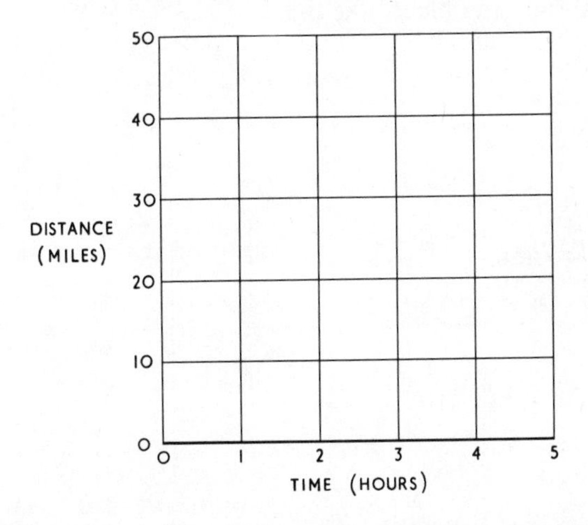

Now although the markings along the graphs represent distance in miles and time in hours, this does not mean that zero time means the actual time at which the universe might have been created and the point marked "1 hour" was one hour after the universe was first created, and so on. Similarly the mark "1 mile" upward does not represent an actual distance of one mile from the center of the universe or some other mythical point. Instead of writing "0, 1, 2, etc. hours" and "0, 1, 2, etc. miles" it would be more accurate to write, "Time at which the experiment started, 1 hour after the experiment started, 2 hours after the experiment started," and so on. Instead of writing "0, 1, 2, etc. miles" it would be more accurate to write "Position of the object at the beginning of the experiment, 1 mile away from the place where the object was at the beginning of the experiment, two miles away (in the same direction) from where the object was . . . ," and so on.

If we realize this clearly, we can see that it would make no difference at all to the graph whether we start the markings on the scales from zero as we have shown in the preceding sketch, or whether we mark them in relation to some actual time and place, like this:

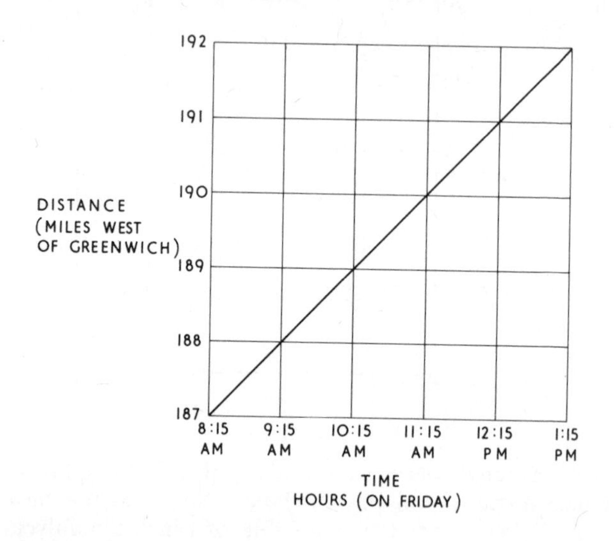

The fourth mark upward would still represent a distance of 4 miles *from* the horizontal base line and the fourth mark along would still represent a time 4 hours *after* the time represented by the vertical base line.

In Chapters 7 and 8 we discussed the slope of a graph at considerable length and pointed out that it was the ratio between the amount up and the amount along, and that this ratio was also the ratio between the *changes* in the two things which were being compared.

In the graph above, for instance, the speed represented by the line is not 187 miles divided by $8\frac{1}{4}$ hours, or 188 miles divided by $9\frac{1}{4}$ hours and so on, but a *change* from 187 to 188 miles (which is one mile) divided by a *change* from $8\frac{1}{4}$ hours to $9\frac{1}{4}$ hours (which is one hour), giving a speed of 1 mile per hour. We can equally say that the slope between the 190 to 191 levels is the change from 190 to 191 miles divided by the change from 11:15 a.m. to 12:15 p.m., which again works out to 1 mile per hour. It is most important that we be clear about this because it is really the foundation on which calculus is built. In fact we could go so far as to say that this is all that differential calculus is, the finding of the ratio between amounts of change in the two things we are comparing.

Suppose now we consider a slightly more complicated relationship—the one between the change of length of a side of a square and the accompanying change of area. We mentioned this relationship before in Chapter 7, page 150. It was pointed out that the area depended on the square of the length of the side. We can make up a simple table, like this:

Length of Side S (feet)	Area of Square $S \times S$ (square feet)
1	$1 \times 1 = 1$
2	$2 \times 2 = 4$
3	$3 \times 3 = 9$

Even before we draw this as a graph it is clear that the relationship between the change in length of the side and the accompanying change in area is not a straightforward one which could be represented by a constant ratio. When the side increases from one unit to two units (1 unit increase) the area increases from 1 unit to 4 units (3 units increase). This gives a ratio of 3 to 1, which is the *average* slope for a side length increasing from 1 to 2 units. But when the side increases from 2 units to 3 units (another increase of 1 unit), the area increases from 4 units to 9 units (an increase of 5 units). This gives a ratio of not 3 to 1 but 5 to 1, which is the *average* slope for a side length increasing from 2 to 3 units. So here we have a situation where the ratio between the change in length and the change in area is not fixed but depends on the actual sizes of the length and area. As explained in Chapter 7, this kind of relationship is found very often in nature: in the relationship between distances and areas, between distance and time in acceleration, between distance and amount of attraction between two magnets, between the amount of light falling on an object and the distance it is away from the source of light. There are literally hundreds of examples of this particular relationship.

So we can see that if we can work out some general relationship we can apply it to all these similar cases without having to do the spadework over again every time.

First of all we will draw a graph showing the relationship between the length of a side (on the horizontal axis) and the corresponding area (on the vertical axis.) The graph looks like this:

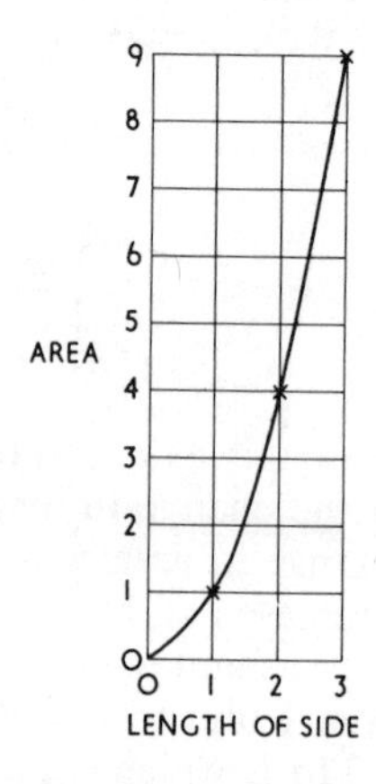

The crosses represent the points listed in the table on the previous page. The problem is to find not an average slope as we have done above but the *actual* slope at any particular point on the graph. The difficulty, as we have seen, is that this is equivalent to dividing 0/0—no change in the vertical direction by no change in the horizontal direction—and getting a real answer.

There are, however, various ingenious ways by which we can get around this difficulty, and here is one of them.

First we begin by choosing, on the curve, a point which is "A" units along the horizontal axis. ("A" represents any actual point we like on the curve, so if we work out an answer for "A" it means we have worked out an answer which will hold good for any point on the curve that we can possibly choose.)

Now since the amount up the vertical axis is always equal to the square of the amount along the horizontal axis for this particular curve, the point which is "A" units along the horizontal axis must also be "A^2" units up the vertical axis, regardless of the value of "A."

We have already found we can't get a value at one point alone so let us choose a second point a little further, say "B" units, along the horizontal axis, "B" being a little larger than "A". This second point will of course be "B^2" up the vertical axis. The relative positions of the two points are shown on the sketch below.

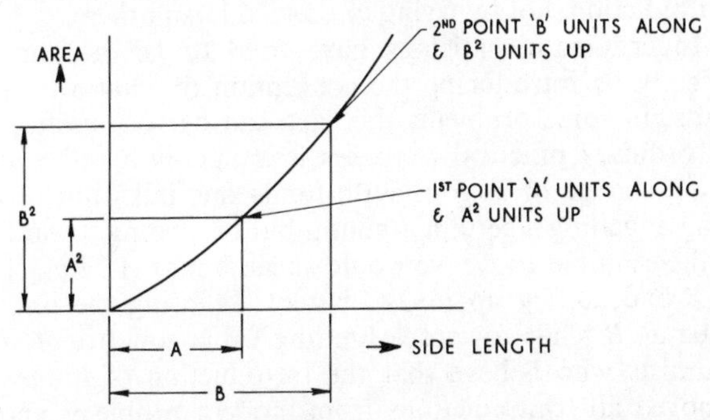

Now we have two points and the average slope of the curve between them is obtained exactly as before. We divide the distance up, (which is B^2-A^2), by the amount along (which is $B-A$). Now it is very easy to prove by simple algebra that B^2-A^2 divided by $B-A$ is equal to $B+A$ or, in other words $(B+A) \times (B-A) = B^2-A^2$. (If you don't know this and don't believe it, then try it substituting numbers for A and B. For instance, 5^2-3^2 divided by $5-3$ is equal to $5+3$. Work it out for yourself.)

So the *average slope*, between these two places, A and B, is equal to $A+B$. This is the same answer as the one we got when we worked out average slopes using actual figures. Between 1 and 2 the slope was $1+2$, or 3; and between 2 and 3 the slope was $2+3$, or 5. Now if we make A and B so close together that they are the same point, we find that A equals B, and the slope at the point A units along the horizontal axis has a value equal to $2 \times A$ exactly.

If we look at this method of working out the slope we will see that we haven't really avoided a contradiction. We pick two points A and B and for the purpose of dividing by $B-A$ we pretend they are *not* the same (otherwise we would be dividing by 0), and then for the purpose of getting the answer we turn around and make them equal after all. But, as in the case of the infinite series, we get an exact answer by this contradictory method. The fact is that reality lies in tacitly acknowledging the existence of the contradiction, not by trying to pretend it isn't there.

Modern mathematicians have tried to get around the difficulty by introducing the conception of what are called limits. In some problems this idea can be very useful, but for ordinary practical purposes it is mainly a rather confusing change of jargon. With limits one talks not of one thing equalling a certain amount, but of tending towards it. In the example above we would speak not of A being equal to B and adding up to $2A$, but of $2A$ being the limiting value as B tends toward a limiting value equal to A. Any optimists who believe that the introduction of limits has removed all contradictions from modern problems should read the following passage, taken from a book explaining the theory of transistors. The book was issued by a world-famous firm of transistor manufacturers, and it is an excellent one. The passage reads as follows:

> "Holes behave like electrons, i.e., like freely moving particles, but are oppositely charged. From the theory

of quantum mechanics, the mass of a hole is slightly different from that of an electron; also they do not move quite as readily."

The devil has restored the status quo with a vengeance!

To return to more practical problems. Since we have found a way of working out the slope at any particular point, let us see what use it can be. In order to save writing the words, "the slope at any particular point in the curve," it is handy to use mathematical symbols. We can hardly write the slope as $\frac{\text{Vertical change}}{\text{Horizontal change}}$ without embarrassment, since there is no change at the point, so the normal practice is to use a lower case "*d*," and we write

$$\frac{d\ \text{vertical}}{d\ \text{horizontal}} \text{ or } \frac{d\,v}{d\,h},$$

which could be translated very roughly to mean "an infinitesimally small bit of vertical change divided by an equivalent bit of horizontal change."

In the graph on page 224 showing the relationship between the area of a square (plotted vertically) and the length of its side (plotted horizontally) we could say that the slope at any particular place was $\frac{d\,A}{d\,l}$ where A is the area and l is the length of the side. And we can further say that this slope, $\frac{d\,A}{d\,l}$ is numerically equal to $2l$, whatever the numerical value of l, the length of the side, may be. So when the length of the side is 1, the slope $\frac{d\,A}{d\,l} = 2$; when the length of the side is 2, the slope $\frac{d\,A}{d\,l} = 4$; when the length of the side is 3, the slope $\frac{d\,A}{d\,l} = 6$, and so on.

Put into plain English, this means that when the length of the side of the square is, say, 3 units long (and the area is $3 \times 3 = 9$ square units) the amount of area is increasing 6 times faster than the amount of length of the side.

This may not seem a very useful piece of information—even if you were putting a concrete border around a tennis court.

But suppose instead of plotting area against the length of a side, we plotted distance vertically against time horizontally. As pointed out earlier, we would then have a graph on which the slope would represent speed (a change of distance divided by a change in time). Here, if D represents units of distance and t represents units of time, we see that if $D = t^2$ then the speed at any instant is the slope, which is $\frac{dD}{dt}$ which equals $2t$. Thus if D is in feet and t in seconds, the speed, $\frac{dD}{dt}$ at 1 second, will be 2 feet per second; at 2 seconds, it will be 4 feet per second; at 3 seconds, 6 feet per second, and so on. From this we can see that the speed will increase by 2 feet per second every second, and that of course is the acceleration.

Another example of the use of differential calculus is the often quoted one of a stone being catapulted straight up into the air. The question is, how high will it rise? Obviously this will depend on its initial upward velocity; but supposing we were told that this was 80 feet per second, where would we go from there?

Before we can begin, there is another piece of information we need. Since the only thing which prevents the stone from travelling upwards indefinitely is the force of gravity, we have to know just what effect this force has. If we make measurements, we find that a free-falling body has, at the end of one second, fallen 16 feet. At the end of the next second it has fallen a total of 64 feet. At the end of three seconds it will have fallen a total distance of 144 feet. Looking at these three amounts we can see that they are 16×1, 16×4, and 16×9. So here we have the distance being equal not to t^2, but to sixteen times t^2. It is still the same basic shape but the slope, instead of being equal to $2t$, is equal to sixteen times $2t$, which is $32t$. Thus by applying differential calculus we can immediately see that the speed

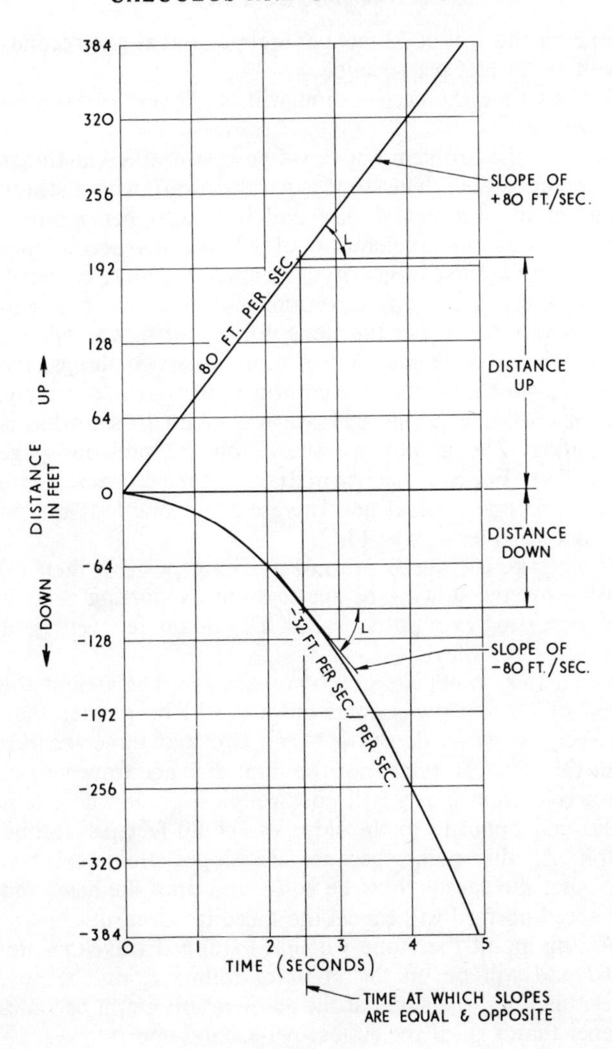
384
320
256
192
128
64
O
-64
-128
-192
-256
-320
-384
DISTANCE IN FEET
UP
DOWN
80 FT. PER SEC.
L
SLOPE OF +80 FT./SEC.
DISTANCE UP
DISTANCE DOWN
-32 FT. PER SEC./PER SEC.
L
SLOPE OF -80 FT./SEC.
O
1
2
3
4
5
TIME (SECONDS)
TIME AT WHICH SLOPES ARE EQUAL & OPPOSITE

at one second will be 32 feet per second, and at two seconds it will be 64 feet per second.

So this time the acceleration will be 32 feet per second per second.

Now in the problem above we have two different things happening at once. The stone is going upward with a steady speed of 80 feet per second and it is also being pulled downward at an acceleration of 32 feet per second per second by the force of gravity. The speed (or more correctly the velocity) in the one direction neutralizes the speed in the other, and whether the stone goes up or down depends on which is the greater. If we plot these two things, the speed upward and the acceleration downward, separately, but on the same graph grid, we will begin to see what is happening. The graphs are shown on the previous page (page 229). (For convenience in drawing the vertical scale we have made one vertical unit represent not one foot but 64 feet. See Chapter 7 page 134.)

If we take the speed upward as being positive then we must show the downward acceleration as starting at zero and increasing in negative value. The minus feet represent distance downward.

From this composite graph we can see that at first the speed of 80 feet per second upward will be greater than the speed downward. But between two and three seconds after the start the *slope* (not the total distance travelled) of the acceleration graph will go through a point where it is equal and opposite to the slope of the 80 feet per second graph. At this point, because the slopes are equal and opposite, the speeds must be equal and opposite also, and the speed upward will cancel the speed downward.

At this point the stone will have stopped travelling upward and will be on the point of falling again, because after this point the slope of the acceleration graph becomes steeper than that of the 80-feet-per-second one.

If you have forgotten all you learned at school about acceleration due to gravity and so on you may by now be feeling rather confused. If so then relax and have a good look at the illustrations below. The first one shows a man leaning forward and dropping a stone over a cliff. Actually there are five separate pictures, something like successive moving picture frames, showing the position of the stone the instant it is dropped and 1, 2, 2½, and 3 seconds later. You can see how, as time goes on, the stone falls faster and faster. Because the distance between the successive pictures is proportional to the time between them, the whole thing is in fact a pictorial graph, and a line can be drawn which represents the position of the stone during any instant over the whole period. Can you see the similarity between this line and the lower half of the graph on page 229?

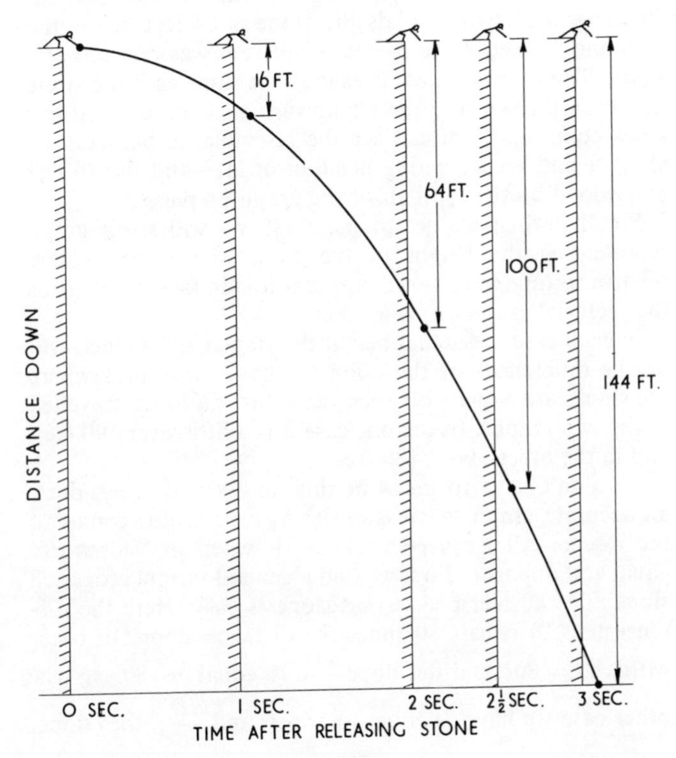

When you feel you understand this then have a look at the sketch on page 233. It is very similar except in this case we have imagined that the person dropping the stone is travelling upward in an elevator at a speed of 80 feet per second. The five successive pictures here show what happens to the man in the elevator and to the stone after he releases it. Notice that, as far as the man in the elevator is concerned, the stone falls away from him *in exactly the same way as before.*

After 1 second it is 16 feet below him, after two seconds, 64 feet below him and so on exactly as before. BUT since in the first second the man has travelled upward a distance of 80 feet and the stone is 16 feet below him, the stone must be $80 - 16 = 64$ feet *above* the place where it was released. At the end of two seconds the stone is 64 feet below the man and 96 feet above the place where it was released and so on. The situation here is exactly the same as if the stone had been thrown into the air upward at a speed of 80 feet per second, and you can see the resemblance between the straight line—representing position of lift—and the 80 feet per second on the top half of the graph on page 229.

Finally, when we get to page 236 we will see a graph representing the height of the stone above the ground relative to time and we will see that it is in fact the same as the pictorial graph on page 233.

In each case the actual height the stone has travelled will be the difference—at the point we have mentioned where the slopes are equal—between the actual distances travelled in the two graphs. In the one case it is a little over 190 feet, and in the other about 100 feet.

We don't have to guess at this however, or even draw an accurate graph to measure it. Again calculus comes to the rescue. All we need to know is when the slopes are equal and opposite. First we find a general formula for each slope. The 80 feet per second slope is easy. Here the distance up (D) equals 80 times the distance along. In other words $D = 80t$, and the slope $\frac{dD}{dt}$ is equal to 80. In the other case we have D equal to $16 \times t^2$ and $\frac{dD}{dt}$, the slope,

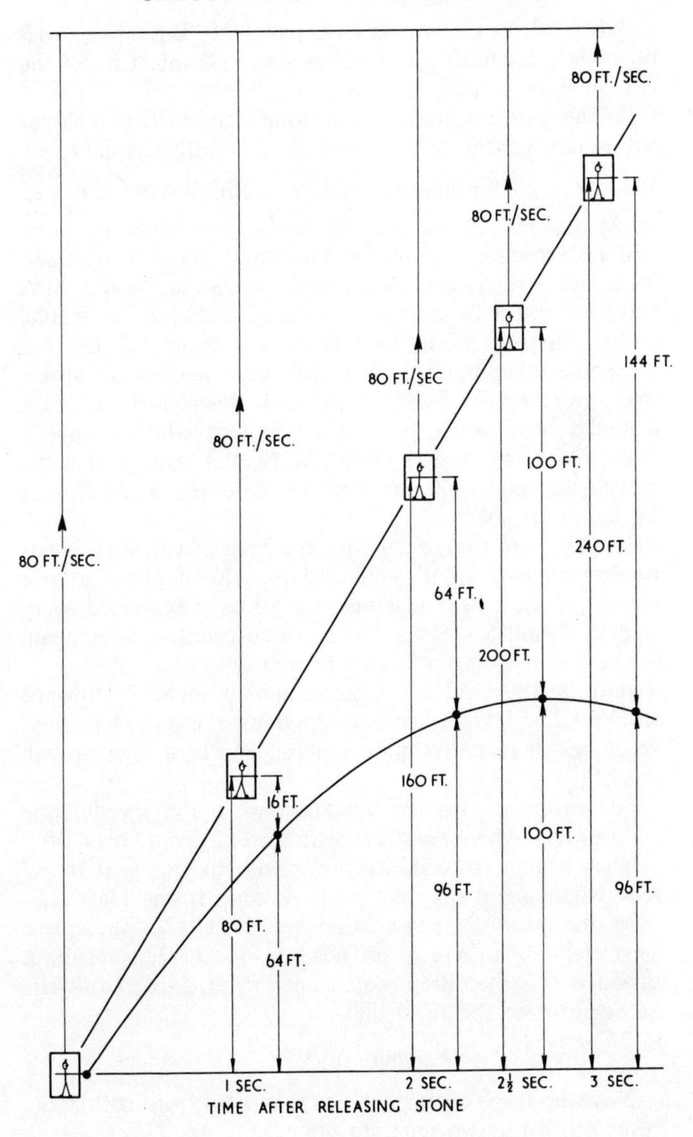
80 FT./SEC.
80 FT./SEC.
80 FT./SEC
80 FT./SEC.
80 FT./SEC.
144 FT.
100 FT.
240 FT.
64 FT.
200 FT.
160 FT.
16 FT.
100 FT.
96 FT.
96 FT.
80 FT.
64 FT.
1 SEC.
2 SEC.
2½ SEC.
3 SEC.
TIME AFTER RELEASING STONE

equal to $32 \times t$ as we saw on page 228. But since this is downward we must put a minus sign in front of it. So the correct slope is $-32 \times t$.

At the topmost point of the stone's flight the two slopes will exactly cancel. In other words, $32 \times t$ will be equal to 80. If $32 \times t = 80$, we find that, dividing both sides by 32, $t = \frac{80}{32}$ or $2\frac{1}{2}$ seconds. So we find that after $2\frac{1}{2}$ seconds the stone will have reached the topmost point of its path. If there had been no acceleration downward, the stone would have travelled up for $2\frac{1}{2}$ seconds at a speed of 80 feet per second and the distance would have been $2\frac{1}{2} \times 80$ or 200 feet.

On the other hand, if there had been no upward speed, the stone would have accelerated downward, and the distance down would have been $16t^2$ feet, which is $16 \times 2\frac{1}{2} \times 2\frac{1}{2}$, which is exactly 100 feet. So the net result is that the stone rises $200-100$, or 100 feet into the air before it begins to fall again.

It may seem that, even with the help of calculus, it has needed quite a lot of time and trouble to arrive at our answer. This, however, is because we have explained every step in detail. So let us do the same problem over again in the way it would normally be done.

Here is the problem: A stone is shot vertically upward at a speed of 80 feet per second against the pull of gravity. When will it reach its highest point, and how high up will this be?

The distance travelled upward due to the speed alone is $80 \times t$ feet, where t is the time in seconds from "blast off."

The distance travelled downward due to gravity is $16 \times t^2$ feet, where again t is the time in seconds from "blast off."

In the previous explanation we dealt with these two separately, but there is no need to do so. The resultant distance D is the difference between them, so we can write an equation to the effect that

$$D = 80t - 16t^2$$

Now the slope of this composite quantity will still be the resultant of the two separate slopes.

So $\frac{dD}{dt} = 80-32t$.

At the point we are interested in, where the stone has stopped going up and hasn't yet started coming down, the slope will be zero (midway between the positive slope indicating a rise and the negative slope indicating a fall).

Thus, at the point we want $\frac{dD}{dt} = 0$

which means that $80-32t = 0$

which means that $\frac{80}{32} = t$

which means that $t = 2\frac{1}{2}$ seconds.

Since D (the distance) $= 80t-16t^2$

we find that when $t = 2\frac{1}{2}$,

$$\text{then } D = (80 \times 2\tfrac{1}{2})-(16 \times 2\tfrac{1}{2} \times 2\tfrac{1}{2})$$
$$= 200-100$$
$$= 100 \text{ feet}$$

So, by using differential calculus and simple arithmetic, we can solve quite difficult and complicated problems.

You may be wondering what the graph of $D = 80t-16t^2$ looks like. We can easily plot it if we find a few values for D, when $t =$ specific numbers such as 0, 1, 2, 3, etc.

Let us make a table of values of D.

When $t =$	Then $D = 80t-16t^2 =$	
0	$(80 \times 0)-(16 \times 0^2) =$	0 feet
1	$(80 \times 1)-(16 \times 1^2) =$	64 feet
2	$(80 \times 2)-(16 \times 2^2) =$	96 feet
3	$(80 \times 3)-(16 \times 3^2) =$	96 feet
4	$(80 \times 4)-(16 \times 4^2) =$	64 feet
5	$(80 \times 5)-(16 \times 5^2) =$	0 feet
6	$(80 \times 6)-(16 \times 6^2) =$	−96 feet

It isn't much use going any further, because if 0 feet represents ground level the stone will have fallen back to the ground in 5 seconds and a second later, if it had not been stopped, would have been 96 feet below ground level, and still falling. So our graph will look like this, if for convenience we make one unit of the vertical scale represent 20 feet.

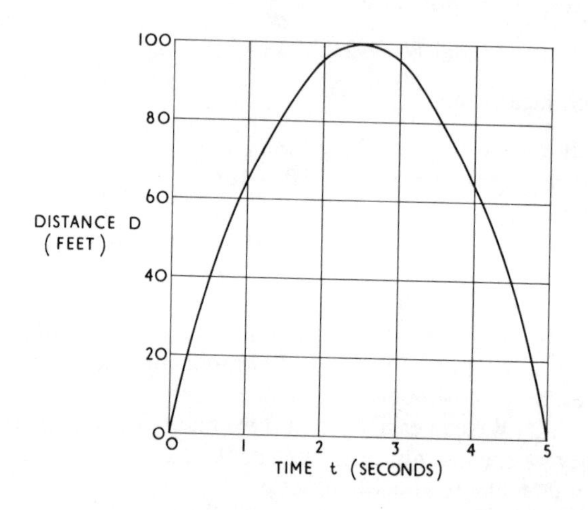

The rising and falling parts are symmetrical and if we look carefully, we will see that each half has the same basic "D is proportional to t^2" kind of shape. The graph also bears out the results we get by using differential calculus. We can see that the maximum height (where the slope is 0) is about 100 feet, and the stone gets there about $2\frac{1}{2}$ seconds after "blast off." But while the graph only gives approximate answers, depending on the accuracy with which it is drawn, differential calculus gives exact answers without plotting points or drawing graphs at all.

There is one final point of interest. By differential calculus we can find that the slope, $\frac{dD}{dt}$, of the graph, $D = 80t - 16t^2$ is equal to $80 - 32t$. Now we can plot this just as we can plot any other equation, and we can make a table fo values of t. Like this:

When $t =$	Then $\frac{dD}{dt}$ (the slope of $D = 80t - 16t^2$) $= 80 - 32t$, which $=$
0	$80 - 0 = 80$
1	$80 - 32 = 48$
2	$80 - 64 = 16$
3	$80 - 96 = -16$

The graph itself would look like this:

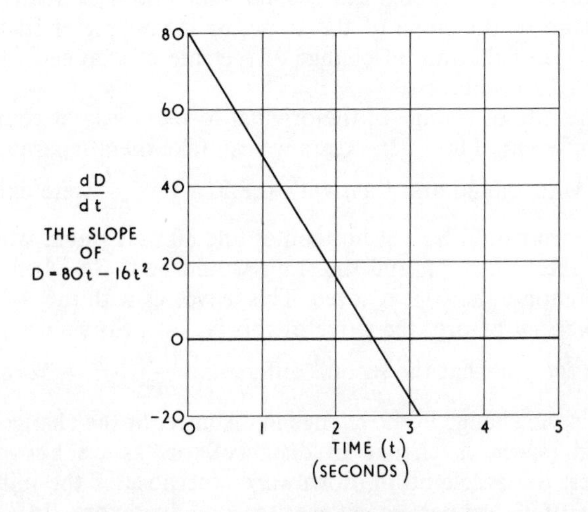

This, of course, only bears out what our calculations showed. The graph represents the slope, which is, in fact, the actual speed of the stone upwards (positive) or downwards (negative) at any period after "blast off." Here again we see that at $2\frac{1}{2}$ seconds the speed is 0, it is neither going up nor down. Before this it was going up (80 feet per second at blast off, 48 feet per second one second later, 16 feet per second a second later still, and one second after that it had reversed direction and was falling at a speed of 16 feet per second).

We can not only plot our new formula $\frac{dD}{dt} = 80-32t$ but, although it may seem surprising at first sight, we can get a differential of it by applying the same basic rules that we applied to the original formula. When this is done it is called the second differential and the symbol for it is written $\frac{d^2D}{dt^2}$. This is sheer jargon designed to confuse students and the little "2s" don't mean that either the *d* or the *t* is squared. It's just one of those things we have to live with. But whatever the symbol, the second differential is really a measure of the slope of the slope, or if you prefer it, the measure of the rate of change of the rate of change of the original formula.

The rate of change of the original formula was of course the differential $80-32t$. Again we can take them separately, first with the 80 and then with the $32t$. If $\frac{d^2D}{dt^2}$ were equal to 80 we would have a horizontal line 80 units high, whatever the value of t, and the slope would be 0. So when we differentiate the 80 is ignored. This leaves us with the $-32t$. As we saw before, the slope of this is -32. So we are left with the fact that the second differential $\frac{d^2t}{dt^2} = -32$ and this is the change in the change in distance, or the change in speed (speed is change in distance), or, as we know it better, the acceleration, downward (because of the minus sign), of 32 feet per second every second due to gravity.

Although the above may be an excellent example of the use of differential calculus, it may not be of any great practical interest unless you are a spaceman or an artillery sergeant. But calculus can be just as useful when applied to problems dealing with change as it can when applied to problems concerning actual physical movement.

Suppose you are thinking of buying a job lot of 100 television sets with the idea of making a living by renting them out. After investigating thoroughly you find two basic facts. First, whether they were rented out or not, each set would cost about $3.50 a month in maintenance, insurance, and deterioration. So you would have to get $350.00 a month for the 100 sets before you made any clear profit.

Secondly, you found that nobody wanted sets at a rental of $12.00 a month, but for every dollar you lowered the rental you could get ten more customers. In other words at $12.00, no rentals; at $11.00, ten customers; at $10.00, twenty customers; and so on until if you came down to $2.00 you could rent out the whole 100 sets.

Before you invested your life savings in the 100 television sets you would want to know what kind of income you could expect. Obviously this would depend on the rental you charged. If you tried to charge $12.00 a month, you wouldn't rent out any at all. On the other hand, if you charged $2.00 you would rent out all the sets but wouldn't be getting enough to pay for the maintenance, which was $3.50 per set. One way you would be down $350.00 per month; the other way you would be down $150.00 per month. Clearly, if there was any profit to be made you would have to fix a price less than $12.00 and more than $2.00, in fact, more than $3.50 because you have to cover the maintenance on the set.

At this stage you will begin to realize that dreams of an income of $150.00 a month are distinctly overoptimistic. Suppose you were actually faced with the problem. What rental do you think would show the most profit? How much clear profit would you expect to get from your 100 sets? $50.00? $75.00? $100.00? Would you take the proposition on at all? Think it over for a minute before you read on, and find if your decision would have landed you in Easy Street or in the bankruptcy court. . . .

Although it may sound foolish, in real-life problems involving calculus, the difficulty is not in applying calculus to a formula but in finding a formula to which calculus can be applied. Once the formula is found the rest is easy. A good way to start is by setting down any basic facts we can think of. For instance, we know that the money coming in each month, that is, income, equals the number of sets rented out multiplied by the monthly rental of each set. We could shorten this statement by writing, Income = Number × Rental. We could even shorten it to $I = N \times R$. It means exactly the same thing.

We also know that the profit we may make is the income we get less the expenditure—in this case for maintenance. So we have another formula: Profit = Income − Maintenance. We can shorten this to $P = I - M$.

Since we already know $I = N \times R$, we can substitute $N \times R$ in place of I and find that $P = N \times R - M$.

One rule for formulas is never to write a symbol when we can put a simple number in its place. We know we have 100 sets and the maintenance on each is $3.50 a month so the Maintenance, M, will be 350. (All the amounts, income, rental, profit, maintenance, etc., are in dollars per month.)

So instead of $P = N \times R - M$, we can now write

$$P = N \times R - 350$$

The only snag now is the rental charge, R. We could work it out if we knew the number of sets rented out, and we could work out the number of sets rented if we knew the rental. But we don't know that either. What we do know, however, is the relationship between the number of sets rented out and R the rental. For every dollar the rental charge is lowered below $12.00 ten sets can be rented out. So if we think a little we can see that the number of sets rented out is equal to ten times the difference between $12.00 and the rental charge. For instance if the rental charge is $9.00, the difference is $12-9$ which is 3 and the number of sets rented out is 10×3, which is 30.

So we can write number of sets rented out = 10 times the difference between 12 and the rental, or ten times 12 − rental.

We can shorten this to $N = 10 \times (12-R)$ and, removing the brackets, this comes to $N = 120 - 10 \times R$.

So now we can get rid of N in the equation above. Instead of writing $P = N \times R - 350$ we can write

$$P = (120 - 10 \times R) \times R - 350$$

Which, when we again remove the brackets, gives us

$$P = 120 \times R - 10 \times R^2 - 350$$

And at last we have our formula for our profit.

After all the trouble in working out the formula the actual differentiation follows exactly the same principles as before. The slope $\frac{dP}{dR}$ of $120 \times R$ is 120. The slope $\frac{dP}{dR}$ of $10 \times R^2$ is $10 \times 2 \times R$ which is $20 \times R$, and the 350 being a plain number and having no slope, disappears. So, for the total formula, the slope $\frac{dP}{dR} = 120 - 20 \times R$.

What does this slope $\frac{dP}{dR}$ really mean? P represents the profit and R represents the rental charged. So the slope $\frac{dP}{dR}$ really represents the actual numerical relationship between the change in rental çharge and the change in profit, which is essentially what we want to know.

You will remember that in the previous example the stone first rose and then fell again, and there was a maximum point in between. Since in the present case there is a loss at both ends of the range of possible rentals, the profit, if there is any, must rise toward the middle of the range and then fall again. If this is so we will have a turning point of zero slope, where $\frac{dP}{dR}$, which is $120 - 20 \times \mathrm{R}$, is zero.

Now if $120 - 20 \times R = 0$, it follows that $120 = 20 \times R$, and R must equal 6. So the differential shows us that the point of maximum profit comes when the rental charge is set at \$6.00.

But before we madly charge in to buy the 100 television sets, let us find what the actual amount of the monthly profit is. We can do this by substituting 6 in place of R in our main formula, namely $P = 120 \times R - 10 \times R^2 - 350$.

Putting 6 for R, $P = 120 \times 6 - 10 \times 6^2 - 350$.

That is $P = 720-360-350$, or $P = 720-710$, or $P = 10$.

So our 100 television sets are going to bring in the handsome total profit, after all expenses are paid, of \$10.00, yes *ten* dollars, not one or two hundred dollars, per month—little over 10¢ per month per set! Aren't you glad you worked the profit out by calculus, instead of finding it out the hard way, with a hundred television sets on your hands?

Perhaps you may not believe this answer could possibly be true. In this case, it is fairly easy to check. Work out the profit for each rental charge. For instance, at \$10.00 there will be 20 customers, which gives an income of \$200.00. The profit will be income less maintenance, which is \$200.00 less \$350.00, which is not a profit, but a loss of \$150.00 per month. And so on. Work it out for yourself.

One final word of caution about formulae. Like everything else in mathematics, they must be used with care and common sense. The formula above gives correct results for all rental charges from \$12.00, where no sets are rented out, to \$2.00, where the whole 100 will be rented out. But if you used a rental charge of, say, \$13.00 in the formula, the formula would assume you are renting -10 sets and calculate your profits (or losses) on that basis. As if things weren't bad enough already! So always remember that formulas can't think and haven't any common sense—that is where you are supposed to come in.

These examples show, in a very simple way, the kind of thing you can do with differential calculus. It is, of course, most useful in situations where working the problem out by trial and error would be so difficult as to be practically impossible.

By successive differentiation we can find the rate of change of the rate of change of the rate of change, etc., stripping change, as it were, of its layers, one after the other.

For instance, the graph on page 219 was a picture of an acceleration of 10 miles per hour per hour. It looked like this:

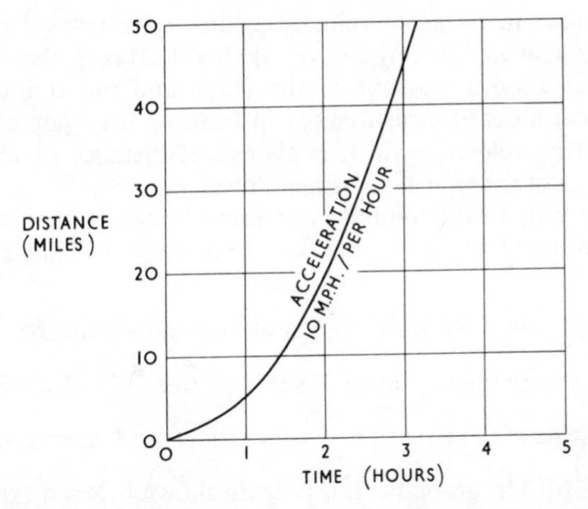

Now the point to notice is that the distance is *not* equal to $10\times t^2$, but to $5\times t^2$, for an acceleration of 10 miles per hour per hour. As explained before, the speed is 10 m.p.h. only at the *end* of the first hour, and the actual distance travelled in the hour is 5 miles. So the formula for the graph is $D = 5\times t^2$. The differential, $\frac{dD}{dt}$, of a graph $D = t^2$ is $2\times t$, so in our case, where $D = 5\times t^2$, then $\frac{dD}{dt} = 5\times 2\times t$ or $10\times t$. The graph of this is shown below on the left.

We can see that it is exactly the same graph as the one on the right where we plot our acceleration of 10 miles per hour

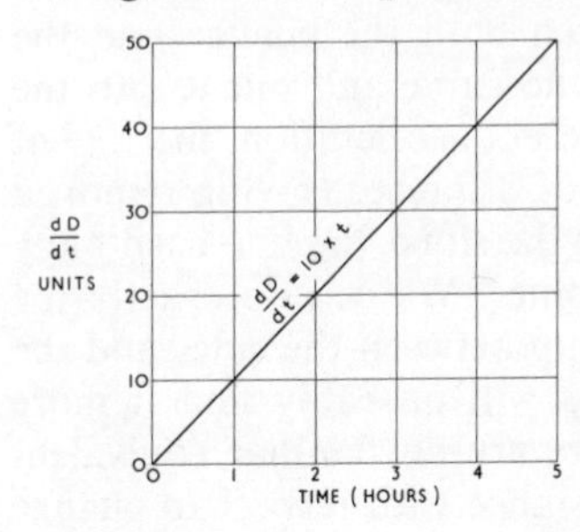

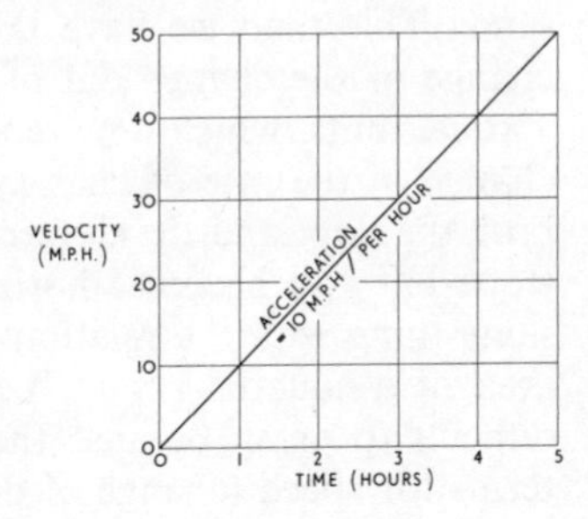

per hour in terms of velocity against time instead of distance against time. We have, as it were, taken the rate of change of distance out of the curve and put it into our vertical units of measurement instead, so that they become speed or velocity (which is change of distance divided by change of time) instead of plain distance.

Now if we differentiate our formula again, that is, find the slope of the slope, or the rate of change of the rate of change, the differential of the differential should be $\dfrac{d\,\frac{dD}{dt}}{dt}$, but as explained on page 238 it is written $\dfrac{d^2D}{dt^2}$. Anyway it is the slope of the equation $\dfrac{dD}{dt} = 10\times t$ and works out to plain 10. The graph for this is again shown below on the left.

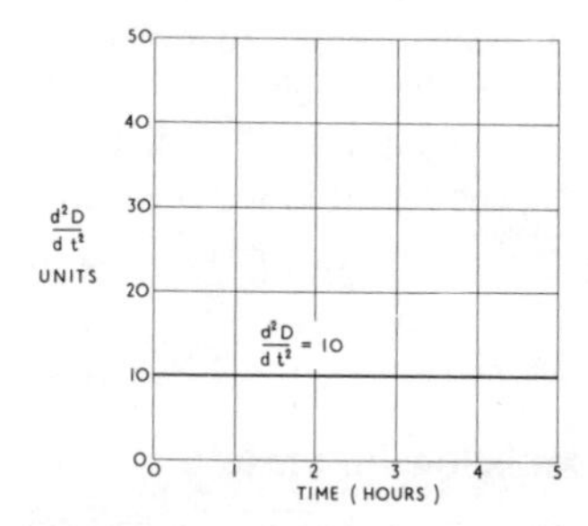

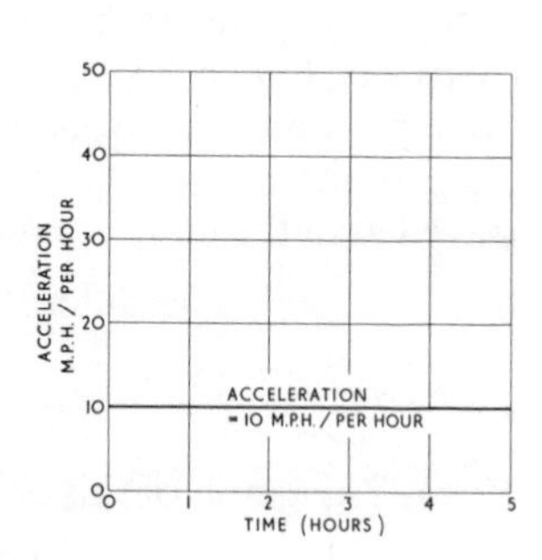

Again we can see that the two graphs are exactly the same. This time we have taken both the change and the change of the change out of the curve and put it into the vertical units, which now represent acceleration, the rate of change of the rate of change of distance. Having removed both the slope and the slope of the slope, the line hasn't any slope left and becomes horizontal. We can do exactly the same thing with the relationship between the sides and the area of a square. Try it. You will probably find it more difficult to grasp because there are no familiar equivalent terms for speed (change of distance with respect to change in time) and acceleration (change or speed with respect to change in time).

Instead you will have to try to visualize, first, change of area with respect to change of length of the sides and, second, change in the change of area with respect to change of length of the sides. If you can do this easily you have a pretty good imagination.

Summing up we can say that differential calculus is the study of change, of rates of change, of the rate of rates of change, and so on. It deals with every conceivable kind of change: changes in speed, changes in direction, changes in size, changes in shape, changes in intensity, changes in stress, and changes in pressure. One can even apply it to changes in social conditions. The only problem here is that we usually have opinions rather than facts to work with!

Before we leave differential calculus there are two ratios which are worth special mention because they give rather unexpected results when differentiated.

One of them is the sine wave and the other is the exponential, or natural number "*e*."

On page 187 we showed a picture of a sine wave and it looked like this:

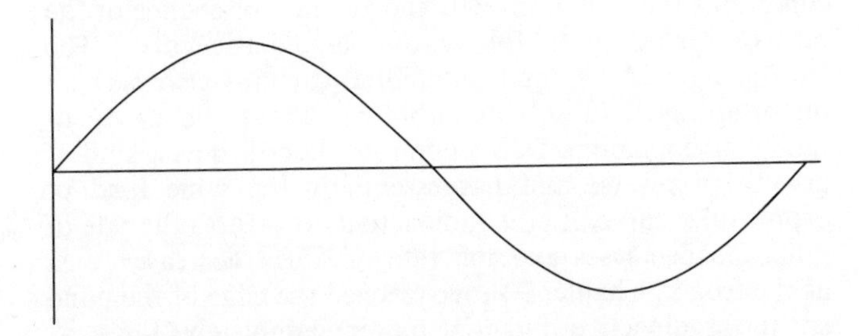

The importance of the sine wave lies in the fact that practically every kind of natural oscillating (or backward-and-forward) movement can be represented graphically by a sine wave-form. The waves in the sea, the swing of a pendulum, radio waves, sound waves, and light waves; the vibration of a tuning fork; all these and many others can be represented by one sine wave or a combination of them.

From a mathematical point of view the peculiar thing about a sine wave is that, when it is differentiated, the answer is what is essentially another sine wave. In other words the rate at which a sine wave changes, and the rate at which the change changes—and so on ad infinitum—of a sine wave is essentially itself. A sine-wave movement appears to be the most "stable" and "unchanging" state of change that exists and this could explain why it is so frequently found in nature. It is, as it were, the "lowest level" of vibratory motion, and all backward-and-forward motions tend to gravitate toward it.

The other ratio, the natural number "e," also has the unusual characteristic that its differential is essentially the same as the original. If we think a little we can see reasons why this should be so. In Chapter 9 we explained how, if a basic rate of growth remained constant, then the actual *amount* of growth taking place at any instant would depend on the actual amount of material at that instant. And this amount in turn depended on the rate of growth. Under these conditions the rate of growth and the rate of change of the rate of growth both follow the same basic curve. The natural number "e" and differential calculus also play an important part in solving problems about the decay of radioactive material. Decay could be described as a kind of growth in reverse and has essentially the same kind of exponential curve. As the radioactivity decreases the rate of radiation decreases and the rate of decay decreases. Just as the frog in Chapter 9 never reached the edge of the pond so the radioactive material never completely loses its activity—or, if you prefer it, only loses its activity after an infinite period. At the same time we have the contradiction that some materials are more radioactive than others.

Imagine that we have a second frog making for the edge of the pond in exactly the same manner as the first but making two jumps for every one jump made by the first frog. Neither of them would ever get there, but it is equally obvious that the second frog is getting nowhere at a much faster rate than the first one! Certainly it will pass the half-way mark before the first one does.

It is exactly this reasoning which scientists apply to radioactivity. The substance which decays more rapidly will lose half of its total radiation in less time than the element which decays more slowly. And that is why we talk of the "half life" of radioactive material. A material which has a half life of one year will lose half its activity during the first twelve months, then *half* of the *remaining* activity during the next twelve months, and so on.

Many people imagine that a radioactive element with a half life of one year will lose all its activity at the end of the second year. Nothing could be further from the truth. It may however be consoling to know that with the aid of calculus it is possible to work out just how much radio-activity will be left at any time up to an infinite number of years.